T0254659

SpringerBriefs in Applied Sciences and Technology

SpringerBriefs present concise summaries of cutting-edge research and practical applications across a wide spectrum of fields. Featuring compact volumes of 50 to 125 pages, the series covers a range of content from professional to academic.

Typical publications can be:

- A timely report of state-of-the art methods
- An introduction to or a manual for the application of mathematical or computer techniques
- A bridge between new research results, as published in journal articles
- A snapshot of a hot or emerging topic
- An in-depth case study
- A presentation of core concepts that students must understand in order to make independent contributions

SpringerBriefs are characterized by fast, global electronic dissemination, standard publishing contracts, standardized manuscript preparation and formatting guidelines, and expedited production schedules.

On the one hand, **SpringerBriefs in Applied Sciences and Technology** are devoted to the publication of fundamentals and applications within the different classical engineering disciplines as well as in interdisciplinary fields that recently emerged between these areas. On the other hand, as the boundary separating fundamental research and applied technology is more and more dissolving, this series is particularly open to trans-disciplinary topics between fundamental science and engineering.

Indexed by EI-Compendex, SCOPUS and Springerlink.

More information about this series at http://www.springer.com/series/8884

Mohammed Dib

Automatic Speech Recognition of Arabic Phonemes with Neural Networks

A Contrastive Study of Arabic and English

Mohammed Dib
University of Tlemcen
Laboratory of Automatic Treatment of the Arabic Language
Tlemcen, Algeria

ISSN 2191-530X ISSN 2191-5318 (electronic)
SpringerBriefs in Applied Sciences and Technology
ISBN 978-3-319-97709-6 ISBN 978-3-319-97710-2 (eBook)
https://doi.org/10.1007/978-3-319-97710-2

Library of Congress Control Number: 2018952916

This Springer imprint is published by the registered company Springer Nature Switzerland AG
The registered company address is: Gewerbestrasse 11, 6330 Cham, Switzerland

To my parents, my wife and my three children: Kheireddine, Rayane and Anness. To my brothers, Abdassalam and AbdelAziz, and to my sisters, Fatima, Aouicha and Zina.

Acknowledgments

I would like to thank Professor Sidi Mohamed Riteri, the Director of the Laboratory of Automatic Treatment of the Arabic Language, for expanding my understanding of applied linguistics, inspiring me to write this book and supplying me with all the necessary tools in his laboratory.

Many thanks go to Pr. Andrew Charles Breeze from the University of Navara and Pr. Dendane Zoubir from the University of Abou Bakr Belaid Tlemcen who have corrected the manuscript with the eyes of an eagle. I thank also Dr. Debbat Fatima of the Department of Electronics at the University of Mascara for providing me with techniques of signal processing and Chatti Boubeker for his help.

I would like to express my deep gratitude to several colleagues—Addou sid Ahmed, teacher of English at the University of Tlemcen; Benammeur Said, teacher in the translation department of the University of Tlemcen; Bensalah Mohamed, teacher of physics; and Abderrahim Amine, Doctor in informatics—for their invaluable assistance.

Contents

About the Author

Mohammed Dib is a Doctor (Maitre de conference A) at the University of Mascara in the English department. He studied at the University of Tlemcen, where he received his license in English, his Magister, and his doctorate in applied linguistics (English-Arabic). His area of interest is contrastive studies in English and Arabic phonetics, phonology, signal processing, speech recognition, and automatic translation. Dr. Dib performed the research for this volume at the Laboratory of Automatic Treatment of the Arabic Language at the University of Tlemcen.

Chapter 1
Introduction

1.1 Contrastive Linguistics

Contrastive linguistics is a field of linguistics which aims to compare linguistic systems of two or more languages in order to ease the task of teaching, learning and translation processes. It has a lot of concerns with teaching problems and therefore tries to provide problem-solving. It provides teaching programmes, on the one hand, and studies the system of each language (syntactic level, phonetic level, phonological level and morphological level) to help in translation on the other. Contrast can be done at several levels; in the syntactic semantic field, for instance, contrastive linguistics works according to universalities, i.e. to delimit how to realize a universal category in contrasted languages. In phonology, however, it deals with phonological characteristics and shows functions of this latter in languages to be compared, i.e. theoretical contrastive study is an independent study; it does not deal with a particular element that exists in language (A). However, it does deal with how a universal category (x) is realized in language (A) and (B). Contrastive linguistic studies, therefore, do not travel from A to B but rather from X to A and X to B (Dresher 2009, p. 1).

Contrastive linguistics appeared in the United States with the help of the Center for Applied Linguistics in Washington, so as to teach German, Spanish, English and Italian. Today contrastive studies have spread out in Europe to make contrast in English, Polish, Finnish, Arabic and German. In the 1950s two books were published; the first, written by Wenreich in 1953, concerns languages in contact; the second, however, was written by Lado (1957) who was considered the founder of contrastive linguistics. He suggests in his book an approach which not only treats grammatical problems but also deals with phonetic and lexical levels whose study requires understanding of cultures (González et al. 2008, p. 205). Lado and Fries are considered the pioneers of contrastive analysis whose works have contributed tremendously to the success of contrastive linguistics during the 1950s and 1960s in

M. Dib, *Automatic Speech Recognition of Arabic Phonemes with Neural Networks*, SpringerBriefs in Applied Sciences and Technology,
https://doi.org/10.1007/978-3-319-97710-2_1

the United States. American and European universities were very interested in contrastive linguistics for the purpose of teaching languages; later on research centres of language teaching and linguists' associations over the world became also interested in this discipline, which gained a prominent place in international conferences of applied linguistics (Hijazi 2013, p. 40).

Contrastive analysis is a technique of contrastive linguistics. It can be defined as a systematic comparison of linguistic features selected from two or more languages to present them to teachers, teaching from a text book which might contribute to teaching materials, course programmes and developing teaching techniques (Fisiak and Mikiewicz 1981, p. 195). Furthermore, contrastive analysis can display similarities and differences of languages clearly. It can be studied with different ways depending on the theory adopted by the researcher, yet the results are different.

Contrastive linguistics is divided into two types: theoretical and practical. The former provides more explanations regarding similarities and dissimilarities that exist between languages and therefore presents an appropriate model for comparison, delimits how to compare elements to each other and shows elements to be compared. It uses concepts such as agreement, equivalence and relation. The latter is a part of applied linguistics; it provides a framework to compare languages, i.e. it selects necessary information for a specific goal such as teaching, dual analysis and translation (Fisiak and Mikiewicz 1981). Moreover, it aims to know how a universal category, let's say X, is realized in language A and how it is viewed in language B, to investigate possible results in application and to know difficulties in another language, for example, when a particular element in the surface structure is absent and interference takes place. Not only are differences taken into consideration, but also similarities are noticed, so the teacher should display them in order not to let the learner think about them and try to build inappropriate forms (Fisiak and Mikiewicz 1981).

In order to describe and compare a set of things, one should possess some tools of measurement which can be applied equally, to compare the weather of different regions, for example, a gauge to measure rainfall and an anemometer to measure the speed of winds are necessary, so that measurement should be made equally between things to have a better result. Contrastive linguistics has features similar to those used in measuring weather in different regions (González et al. 2008, p. 207).

As for the contrastive method, it was noticed that in groups learning English, learners who have other mother tongues experience different problems in learning the second language, i.e. English. In other words, the mother tongue has an impact called the negative transfer on the learning of foreign languages. A very careful comparison can show problems learners face during the course, thus anticipating mistakes that they can make; thereby delicate problems appear in sharp differences.

Contrastive linguistics has two goals: theoretical and practical. The theoretical goal is concerned with the rise of knowledge in the field of linguistics, while the practical goal deals with language teaching and presenting teaching material (Chitoran 1972, p. 37).

To sum up, contrastive linguistics supplies foreign-language teachers with material which can be found only by coincidence. Grawhil advocates the comparison of equivalent portions of two languages is made for the purpose of isolating the problems that a speaker of one language will have in acquiring the other (McGraw-Hill 1966, p. 287).

1.2 Computing Contrast

The notion of contrast has been considered as a pivotal point in linguistics since the advent of Ferdinand De Saussure, who argues that there are only differences in language. He adds that the sound in language is not more important in itself than the differences which can distinguish words from each other. In other words, the phoneme is not defined by its characteristics but rather by the characteristics that are found in the other phonemes (Dresher 2009, p. 1). Contrast can be found at several levels, from the phonetic level to the pragmatic one. At the fundamental level, it can show the difference between two sounds. For instance, in English, there are two sounds, long /i/ and short/ i/, which can distinguish between *feel* and *fill*; simply put, this contrast recurs in several words such as *cheap*, *chip*, *seat*, *sit*, *seen*, *sin*, *meal*, *mill*, *reed* and *rid*, so the important thing in the word is not the sound itself, but differences which can help to distinguish between words.

Acquiring phonological contrast is one of the difficult tasks with regard to learners and machines as well, so determining contrast is a basic characteristic for phonological description and a fundamental condition for analysis. In this respect, phonologists have tried to explore contrastive characteristics and the manner to study them. Each one has come out with an idea. Jakobson argues that the core of phonology is a set of distinctive features used by all languages to build contrast (Lass 1984, p. 75). Dresher, however, defines a minimal pair as a contrast between two sounds that exist in the inventory of language, although distinctive features are found only in one characteristic (Dresher 2009, p. 20). As regards computing contrast, it should be noted that what has been said earlier should be formalized. The reason is that the target is the machine. Many linguists have discussed this problem, among them Noam Chomsky, Halle, Andre Martinet and Roman Jakobson. Chomsky proposes integers for the classification of binary features. For instance, vowel height is explained as follows:

Sound	i:	e	ɛ	ɛ:	a	ɑ:	u	u	o	ɔ
Heigh	+	+	+	–	–	–	+	–	+	–
Mid	–	+	+	+	–	–	–	–	+	–
Back	–	–	–	–	–	–	+	+	+	+
Long	+	–	–	+	+	–	+	–	–	+

André Martinet, however, suggests isolating contrastive features in academic French consonants. To present a fruitful account, the focus should be made only on bilabial plosive consonants/ b,p,m/. Martinet sees that / b/ is contrasted with /p/ in voicing, whereas /m/ is contrasted with /b/ in nasalization. This description can be displayed by binary features isolating the characteristics which can lead to distinguishing between different phonemes. Focusing on voicing and nasalization will have the following:

- Voicing and nasalization
- [+voice] = voiced [−voice] = voiceless
- [+nasal] = nasal
- [−nasal] = nonnasal

/p/ and /b/ differ only in voicing, and if there is no contrast in this feature, it won't be easy to distinguish between them, but/ b/ and /m/ differ in nasalization, which is considered as a basic feature for contrast. Let's circle the contrasted sounds.

Sound	p	b	m
Voicing	⊖	⊕	+
Nasalization	–	⊖	⊕

What about features that are not circled? These features are put aside because they are not in contrast, so that voicing is not considered as a distinctive feature in /m/,/n/ and/ɳ / because not only are they voiced, but they are also nasalized. As a result, computing contrast requires three things: contrast, as a procedure, then extracting similarities and then extracting distinctive features.

1.3 General Introduction

This survey is under the heading ***"Automatic Speech Recognition of Arabic Phonemes with Neural Networks: A Contrastive Study of Arabic and English"***. It is undoubtedly that the thing which spurred me to tackle this work is the sharp deficiencies of the Arabic language as far as computer sciences are concerned, more simply put its deceleration in the scientific field, which may discourage to take a position in the era of globalization. So is this deceleration due to its complicated system or to the apathy of its searchers?

In this research we would justify that:

- Arabic is a flexible language.
- Its phonological system can help easily in automatic speech recognition.
- Its limited number of vowels can help generate an infinite number of syllables.

The main purpose of this survey is to treat the Arabic minimal syllable automatically so as to contribute to the enrichment of the Arabic treasury, expand its use in computational field, and develop it in the field of technology to go on a par with globalization. Thus trace a road map for teachers who find difficulty in teaching

English phonology to Arab speakers throughout a contrastive analysis with selected corpora highlighting the main differences between both languages, both at the phonetic and phonological level.

In so doing, a set of data and different approaches and methods have been exploited. The data used in this work come from different sources. The examples displayed to illustrate phonological aspects in English and Arabic have been extracted from two corpora of journalistic language, where English corpus contains an article entitled *The Man with a Bionic* ***Eye***, while Arabic corpus encompasses the following article:

"سرطان الكلى أساليب علاجية جديدة"

which can be roughly translated into English as "**Kidney Cancer, New Styles of Cure**". The examples to study contrast were selected from both corpora to illustrate the difference between linguistic units in both languages, mainly the weak syllable containing the schwa vowel and syllables involving (h) sound in final position. Contrasts here are studied as isolated linguistic facts (case of pause). In so doing, taxonomic phonemic and spectral analyses were applied. The data base in the last part of the work has been gathered to establish a knowledge base and has been recorded by a number of speakers to acquire a sound reference. The knowledge base has been injected in the neural networks for purpose of sound training and recognition. Results of previous corpus-based studies were also used to ensure the empirical basis of the descriptive and the contrastive claims. The methodology followed in this work is contrastive analysis based on method of transfer, Harris taxonomy and a generative approach, in addition to a neural approach which adheres to spectral analysis and neural network system.

This book comprises four chapters; two of them are dedicated to theoretical works concerning both the Arabic and English languages, and the last two chapters are devoted to practical and spectral analysis. This organization is based on two main things: contrastive analysis and neural approach. Chapter 1 is dedicated to language system, the impact of environment upon language, fluency (***alfasaha***), Arabic sounds, the syllable in the Arabic language, distinctive characteristics of syllable, types of syllables in Arabic, stress in Arabic and intonation in Arabic. Chapter 2 encompasses sound, classification of sounds in English, the nature of the syllable in English, the structure of the syllable in English, weak vowel (schwa), the position of the syllable, syllabic consonant, stress and intonation. Chapter 3 presents a contrastive study in English and Arabic, a chart of Arabic sounds, a chart of English sounds, a corpus entailing the English and Arabic articles, phonetic transcription, a practical survey, contrastive analysis and spectral analysis. Chapter 4 tackles speech recognition, introduction, speech production, spectral analysis, speech recognition, fundamentals of speech recognition, the notion of intelligence, artificial intelligence, programmation, neural networks, fundamental of neural networks, analysis of speech signals, acoustic phonetics, formants, data base, ways of processing, signal processing, the definition of signal, frequency, wave, amplitude, spectrum, physical sound, practical work and a conclusion.

References

Chitoran D (1972) The Romanian English contrastive. In: AILA 3rd Congress, Copenhagen Proceedings, vol 1

Dresher BE (2009) The contrastive hierarchy in phonology, 1st edn. Cambridge University Press, New York

Fisiak J, Mikiewicz A (1981) Contrastive linguistics and the language teacher, 1st edn. Pergamon, Oxford. Reprinted 1985 ISBN 0-08-027230-4

González MÁG, Mackenzie JL, Álvarez EMG (eds) (2008) Current trends in contrastive linguistics functional and cognitive perspectives. John Benjamins, Amsterdam

Lado R (1957) Linguistics across cultures: applied linguistics for language teachers. University of Michigan Press ELT, Ann Arbor

Lass R (1984) Phonology an introduction to basic concept, 1st edn. Cambridge University Press, Cambridge

McGraw-Hill (1966) Trends in language teaching. McGraw-Hill, New York

Arabic Reference

Hijazi MF (2013) *Ilm allugha alarabiya: Madkhal Tarikhi fi Dawii Tutat wa allughat Samiya* (The science of Aarabic language: a comparative, historical prelude in the light of patrimony and Semitic languages, 1st edn. Gharib Library)

Chapter 2
The Arabic Phonological System

2.1 Introduction

There is no doubt that the main function of language is communication and transmitting ideas from the speaker to the listener; therefore "language is a means used by people to understand each other". It is also a general human phenomenon that serves the same functions in different societies. It is made of a set of sounds that fulfil different meanings. From this standpoint, one can see that languages are different from one community to another and every language has its own system which is studied in terms of phonetics, phonology, morphology and semantics. The idea of Firth is a good demonstration of the point. He relates language to society and sees that humans communicate to each other according to different social attitudes using a particular form of style and choosing different types of words. So language is influenced by the social framework that is used within it and is influenced by its elements (Almasdi 1986, p. 176).

2.2 Impact of Environment on Language

The environment is considered the source of the cultural elements that influence human linguistic behaviour. Environment can be divided into the physical nature, that is, beyond the human control and the social environment as well. Analysing language, which is said to be a set of symbols that evoke the physical and the social world where people live, requires the use of the term environment, which includes the physical and social factors. The physical environment entails the geographic characteristics such as topography (coasts, plains, hills and mountains), climate, fall of rain, fauna and flora and mineral sources.

M. Dib, *Automatic Speech Recognition of Arabic Phonemes with Neural Networks*, SpringerBriefs in Applied Sciences and Technology,
https://doi.org/10.1007/978-3-319-97710-2_2

The social environment refers to the different forces which build the lives and ideas of people, such as religion, ethical values and the political system. So the factors affecting the human culture can be classed into the aforementioned environments, which are regarded as two main things that contribute to exhibit the language. For instance, the existence of a particular kind of animal is not sufficient to create a linguistic symbol and be used in a given language; instead, members of the community should know this kind of animal and should have a common interest in it (Sapir 1967, p. 73).

These effects can be applied to language in four levels: vocabulary, phonology, morphology and syntax (Sapir 1967). In this respect, Chomsky as quoted in Radford et al. 2010, p. 4 said: 'Language is a cognitive system which is a part of any normal human being's mental or psychological structure'. He adds that if we assert that Tom is a speaker of English, we are ascribing that Tom's brain is in a certain structure, that is, the word mail in English consists of a consonant, a diphthong and a consonant with regard to the English speaker, in contrast to the word ميل, which means leaning and comprises a consonant, a vowel and two consonants though they are similar in the linguistic realization. From this standpoint, we can say that grammar evokes the behaviour of the speaker, the idea that Chomsky emphasizes by saying that grammar evokes the behaviour of the speaker who can produce and understand an infinite number of grammatical sentences starting from a limited experience (Kristeva 1989, p. 254).

2.3 Alfasaha (Fluency or Eloquency)

In Arabic, fluency stands for clarity. It is a noun used for a single word, speech and speaker as well. We say a fluent word, a fluent speaker and fluent speech. The fluency of the word is characterized by non-strangeness of speech and no misuse of analogy and sounds (Touiji 2001, p. 449).

Cacophony or clashing sounds can be detected by the full understanding of the Arabic speech. The following verse said by the poet Imru al-Qais is a good demonstration of the point.

غدائره مستشزرات إلى العلا .... تظل العقاس من مثني و مرسل(Al Suyuti 1987 p. 37)
which can be translated in English as follows:
Her tresses are elevated to heights
The ponytails are lost among the braids and the curls

The word ***mustachzirat*** is characterized by cacophony and clashing sounds which seem strange with regard to the listener, that is, unusual in terms of speech. This is called the defiance of morphophonemic norms. Fluency in the singular word is the desire for and not disgust in listening, such as the word ***aljarachi*** in the verse said by Al-Mutanabbi.

مبارك الأمم أغر اللقب كريم الجرشي شريف النسب

(Quazouini 1967, p. 39)
Blessed in name and prominent by epithet
Noble of self and honorable of lineage (Learn Arabic.com).

Eloquence in speech should be free from defiance in grammatical norms, clashing words and convolution as well. Defiance of grammatical norms is due to the defiance of analogy, such as the following verse which is attributed to Ibn Djinni.

وقبر حرب بمكان قفر و ليس قرب قبر حرب قبر

The grave of Harb is in a desolate place
And there is no grave near the grave of Harb

2.3.1 Arabic Sounds

The number of Arabic sounds is 29; the (***hamza***) comes first; it is the equivalence of the glottal stop in English and can be written on the three long vowel letters ***alif***, ***yaa*** and ***waw***. Abu alAbass Almubarad, one of the Arab grammarians, however, declares that the number of Arabic sounds is only 28, putting the (***hamza***) aside; his argumentation is that the (***hamza***) has no form, for sometimes it is written ***waw***, other times ***alif*** and other times ***yaa***; therefore it is not included with sounds whose form is independent and known; it is found in speech but shown in written form by symbols, because it has no image (Zamakhshari 2001, p. 198).

Sybawayh argues that Arabic sounds can be divided into fundamentals and are 28 including the ***alif*** and secondary which some of them are acceptable like discreet /n/ and ***hamza*** in between, inclined ***alif***, [ʃ] which resembles [ʒ], [ṣ] resembling [z] and emphatic ***alif***, and some of them are unacceptable such as [k] whose place of articulation is between [k] and [ʒ] and [ṭ] resembling [t].

To sum up, the number of Arabic sounds according to Arab grammarians can reach the number of 50, but the reader can ask about the real number of these sounds. The answer is that, phonologically speaking, Arabic sounds are 28, while the rest of the sounds are only allophones.

2.3.2 Sounds Used by Arabs

According to Ibn Yazid, the sounds that are most used by Arabs are ***waw***, ***yaa*** and the ***hamza***, and the sounds less used due to the effort that they require are ***taa***, ***ðal***, ***θaa***, ***ʃin***, ***qaf***, ***haa***, ***ʔajn***, ***nun***, ***lam***, ***raa***, ***baa***, and ***mim***, yet the easily pronounced sounds are those used to build new words, i.e. affixes. He added that degrees of fluency are different. The word is considered heavy or light according to its transition from one sound to another. If the word is triliteral, it will have 12 structures, which are as follows:

- First: Transition from the high place of articulation to the mid to the front (ʔ, d, b)
- Second: Transition from the high place of articulation to the mid to the front (ʔ, r, d)
- Third: Transition from the high place of articulation to the mid to the high)h, m, ʔ)
- Fourth: Transition from high to mid to high (ʔ l, n,)
- Fifth: Transition from front to mid to high (b, d, ʔ)

- Sixth: Transition from front to high to mid (b, ʔ, d,)
- Seven: Transition from front to high to front (f, ʔ, m)
- Eight: Transition from front to mid to front (f, d, m)
- Ninth: Transition from mid to high to front (d, ʔ, m)
- Tenth: Transition from mid to front to high (d, m, ʔ)
- Eleven: Transition from mid to high to mid (n, ʔ, l)
- Twelfth: Transition from mid to front to mid (n, m, l) (Al Suyuti 1987, p. 198).
- If this occurs, the better and the most used structures are those which shift from high to mid to front, then which shift from mid to front to high and then which shift from high to front to mid (Al Suyuti 1987).
- The [m] sound is similar to [w] sound because they have the same place of articulation, i.e. bilabial; however the [m] sound contains nasality, where the soft palate is raised a little bit so as to allow the air to move freely through both oral and nasal cavity.

2.3.3 The Syllable in Arabic

The point of departure of syllable in Arabic is always a consonant, while its closure is made up of either a vowel (called the open syllable) or with a consonant (called the close syllable) or with two consonants called (long syllable closed). In other words, the syllable in Arabic never starts with a consonant cluster but can end with a consonant cluster. More simply put, Arabic rejects the starting of two connected plosive consonants. Without further ado, here are some examples which clarify the point: the word ***uktub*** is the imperative form of the verb /kataba/, to write whose structure requires the syllable onset [ʔu] in the first position as it is impossible to say ***/ktub/*** because the Arabic phonological system refuses consonant cluster. To prevent syllable initial consonant cluster, a vowel is added as in ***/uktub/***, but as Arabic also prevents vowel initial syllables or words, a ***hamza***-glottal stop is added in initial position. However, the ***hamza*** (called hamzat wasl) is easily elided when the word is preceded by some other word as [ɵummaktub]. The Greek word ***platon*** is the like; it has been reshaped as Aflaton in Arabic, and the French word ***franc*** has changed into Ifrange (Flesh 2007, p. 43 cited in AWEJ V5N4 December 2014).

Overall, the syllable is a connection made by a consonant and a vowel. Roman Jakobson states that the syllable is a group of structures which entails two associated phonemes with different degrees of aperture, one of a smaller degree and the other of a bigger one (Abdelouhab 1984, p. 27 cited in AWEJ V5N4 December 2014).

Amber Crombie, however, declares that speech relies first and foremost upon breathing and the airflow is pulse-like; the muscle contraction caused by the air pressure makes a chest pulse which in turn produces a syllable. The rhythm of chest pulses is the essence of human language (Hanun 1992, p. 65 cited in AWEJ V5N4 December 2014).

Some sources said that the segmentation of speech into syllables goes back to a long time with regard to Arabs, to the period when Arabic was purely an oral language and relied only on the listening process to transfer literature and arts.

Aljahid, one of the Arab grammarians, utilized the term syllabification, meaning segmentation of speech. He said that the sound is the mechanism of speech through which syllabification and connection are made (Aljahid 2010, p. 79 cited in AWEJ V5N4 December 2014). Connected speech encompasses syllables that bear the phonotactics of a particular language. We mentioned earlier that syllables may consist of open and closed ones. Open syllables are those which end with short or long vowel, while closed syllables are those which end with a consonant with an absence of the mark referred to as ***sukun***. The triliteral root verb ***fataha*** contains three open syllables, yet the noun ***fathun*** comprises two closed syllables: ***fat*** and ***hun*** (Annis 1997, p. 162 cited in AWEJ V5N4 December 2014).

2.3.4 Structure of the Arabic Syllable

A syllable is an association of a consonant and a vowel. This combination is made according to the system of any language in constructing its syllable structure. It relies upon the respiratory rhythm. The formation of a minimal syllable in Arabic, for instance, is made up of one consonant followed by a short or a long vowel; in another word, a sequence of two consecutive consonants is rejected apart from the case of pause. The syllable in Arabic never starts with two consonants or a vowel. It is certainly the thing which can explain the use of a linking sound called ***hamzat alwasl*** (glottal stop) used for linking.

2.3.5 Types of Syllables

In all languages, syllables consist of vowels as centres preceded or followed by consonants despite the differences that exist between languages over the location of consonants, but in some cases, syllables may be formed without a vowel. Czech words are a good demonstration of the point ***prno*** and ***Vltava***, where these syllables consist of consonants only.

One point three of the studied languages uses consonants only to form syllables (Hagége 1982, p. 24). German and English are among those languages, but Arabic is excluded from that because its phonological system rejects two consecutive consonants.

The German word abend is pronounced [abant] in careful speech, whereas in connected speech it is pronounced [abnt] or [abmt]; likewise English words ‘bottle’ and ‘button’ are pronounced [bɒtl], [bʌtn]. Syllabic consonant appears because of the deletion of the weak vowel schwa (Roach 1992, p. 106).

There is a controversy about the number of syllables in Arabic. Some linguists argue that there are six types, but others say there are only five. They are as follows:

- ***The minimal syllable***. It is made up of a consonant and a vowel. They may be significant linguistic units if they express prepositions, e.g. bi, fi, li, etc. (cv).
- ***Closed long syllable***. It is composed of a consonant, a vowel and a consonant, e.g. min (from)bal(rather) (cvc).
- ***Open long syllable***. It is made up of a consonant and two vowels, e.g. maa haa (cvv).
- Two forms related to the pause.
- ***Long syllable closed with a consonant***. It consists of a consonant+two vowels +a consonant, e.g. kaan (was) (cvvc).
- ***Long syllable closed with two consonant***. Its structure is made up of a consonant+a vowel+two consonants. e.g. karb fadl (cvcc) (Kaddour 1999, p. 75 cited in AWEJ V5N4 December 2014).

2.4 Stress

Stress results from the production of energy by the organs of speech, and it can be studied from two points of view: production, that is, which organ of speech is involved during the production of syllables and which syllable appears to be stressed with regard to the listener (Roach 1991). It is also defined as an extra energy related to a syllable, in order to show its degree of prominence and an activity made by all speech organs at the same time, that is, a movement of speech organs is noticeable when uttering a stressed syllable; in other words, chest muscles and lungs make an effort; thus vocal folds get closer to each other; hence amplitudes of vibration become greater, and as a result the sound becomes louder and more prominent; this in case of voiced sounds (Annis 1997, p. 172).

Stress in Arabic has a fixed place; therefore it can be governed by a rule and has no impact upon meaning. This can be seen clearly in the French language when trying to investigate the rule of stress which falls most of the time on the first syllable (Alqamati 1986, p. 70). Avicena argues that stress is the process of producing energy consisting in exhaling a big amount of air starting from the diaphragm and pushed out by the chest muscle (Avicenna 1983, p. 72).

In Arabic, stress falls on the long unit which can be either a long vowel or two neighbouring consonants even if they are two different syllables, e.g.:

/شـا/•/ـا/ـا//ه//د/ـــ/ meaningد: he observed / /شاه•//هـ//ــ/ــ/ــد/ // /د/// ه• //شا sha •hada / [ʃaahada] [xaradʒnaa] /خـ•/ـا/ـرجـ/ـا/ـنـا/meaning: we got out

If there is no long unit, the stress falls on the antepenultimate, e.g.

/ فـ•/ـا/ـهـ/ـا/ـمـ/

which stands for he understood (fa• hima) and

(wa• raq) which means page.

/و•// ر// ق/

From these examples one can understand that stress in Arabic is fixed and predictable. It is not a distinctive phonetic unit and has no influence on the meaning. It is acceptable, therefore, to pronounce the following words into two ways:

munta•damun // مـظـ/ـ/ • ـتـ/ـ/منـ/ (organized) or mu•ntadamun / مـــظـــ/ـــ/ــتــ/ـــ/ • منــ/ (Alqamati 1986, p. 70).

As for stress in English, it is called free or movable stress, i.e. it is unpredictable and is a suprasegmental feature which has an impact on morphological structure and the meaning of the word as well (Alqamati 1986). It has, then, a morphological and semantic function. The morphological function may be observed in the word import which works as a verb when the stress falls on the second syllable /ɪmp'ɔ:t/and as a noun when the first syllable is stressed/'ɪmpɔ:t/. As for the semantic function, the word august is a good demonstration of the point. Stressed on the first syllable, the word means month, while stressed on the second syllable, it stands for an elegant man (Roach 1991, p. 50).

So, stress is an activity made at the level of speech organs in the same syllable at the same time. When, for instance, a stressed word is articulated, the most important thing to notice is that all the organs of speech are in movement; in other words, the lungs and chest muscles are squeezed, and vocal cords approach each other so as not to let a big amount of the air flow; therefore frequencies of vibration increase, and consequently the sound becomes prominent; this is as far as voiced sounds are concerned. As regards voiceless sounds, vocal cords are wide apart, so as to allow a big quantity of air to pass (Annis 1997, p. 172).

It is also important to notice that in uttering a stressed syllable, there is an activity of the hard palate, the tongue and the lips, but in case of an unstressed sound, speech organs remain still. The distance between the vocal cords when uttering voiced sounds expands relatively, so the air pressure and amplitude of vibrations become smaller as well.

Speakers of any language tend to press on a specific syllable in order to make it very clear compared to other syllables in the word; this pressure is called stress (Annis 1997).

Languages differ in terms of place of stress. Some languages have predictable stress, that is, the stress is governed by a rule, e.g. Arabic and French, but other languages have unpredictable place of stress, i.e., the stress is very hard to capture, e.g. English.

"The French speaker generally stresses the first syllable in each word"; thus once trying to speak English, he applies stress once more on the first syllable of English words because he is affected by his linguistic habit, the thing which causes ambiguity in the meaning of words, since the meaning of words in English may change due to the place of stress, e.g. (**import**, **augment** (Annis 1997).

There is no evidence which informs us about the place of stress in Arabic, that is, how people used to pronounce words in the first ages. But as for Quranic reading, there is a rule concerning stress which is as follows:

- To know the place of stress in Arabic, we should look at its last syllable, that is, if it is of the fourth or fifth type; it is, therefore, the important syllable which bears stress, and this is only in case of pause. Simply put, stress in Arabic never falls on the last syllable except in case of pause, and the last syllable must be of the fourth or the fifth type, i.e. a consonant+two vowels +a consonant or a consonant +vowel +two consonants. In case of pause, for instance, in the Quranic verse {إِيَّاكَ نَعْبُدُ وإِيَّاكَ نَسْتَعِينْ}, which is roughly translated as ***Thee alone we worship***, ***Thee alone we ask for help*** or the pause in almustaqar in the following Quranic verse{ إلى ربك يومئذ المستقر } translated as ***To your lord that day in the place of permanence***, stress is found in the last two syllables "قر" و "عين".

- If the mentioned syllables do not exist in a word, the stress falls on the penultimate, which should not be of the first type or proceeded by the same type (Annis 1971).

2.5 Compound Vowel

2.5.1 Definition

There are two terms for the compound vowels; the first is ***hiatus*** and concerns two vowels, and this is not acceptable in Arabic, because the Arabic phonotactic system rejects this phenomenon. The second, however, is a sequence of a vowel and a semivowel and goes on a par with the Arabic system.

In Arabic there are two semivowels: the ***waw*** and the ***ya***. If they are preceded by a vowel, they are called falling diphthongs, but if the vowel follows the semivowel, it is, then, called the rising diphthong (Ababna 2000, p. 131).

There is a controversy among Arabic linguists and phonologists concerning the term diphthong in Arabic. Some phonologists admit that all languages contain diphthongs. Vandrice argues that diphthongs exist in all languages; this view was supported by Abdassabur Shahin who sees that the ignorance of diphthongs is due to the writing system which represents only a half of linguistic reality (Annis 1971).

Other linguists argue that the diphthong is a phenomenon which exists in all languages but is a phonetic context that consists of two consecutive vowels or more in a syllable; this is true as far as the system of the European languages and their writing systems are concerned, but it is not acceptable in the Arabic language. Bartil Malberj pointed to the concept of the diphthong, where a semivowel follows the vowel, saying that modern French does not contain diphthongs and **oui**, **oi**, **ui** and **ei** in the following words (**pied nuit**, **fois**) are merely a sequence of a consonant and vowel.

2.5.2 Difference Between Diphthong and Vowel

There is a sharp difference between a diphthong and a vowel. When pronouncing a vowel, the speech organs remain in their place of articulation spending a given moment of time, while in diphthongs we may notice a connection between two pronounced vowels forming one syllable. In fact, it is a synthesis of two sounds which contain an intended glide, so that the speech organs take one position and then move directly to another one (Saaran 1994, p. 154).

As mentioned earlier, the diphthong is a sound which results from a following vowel and a semivowel, whereas the ***waw*** and the ***ya*** are called semivowels. Ibrahim Annis said that the ***waw*** and the ***ya*** must exist to form a diphthong (Saaran 1994). The diphthong may appear as a result of two connected words.

2.6 Intonation in Arabic

Before embarking on any discussion of the types of intonation, it is very important to mention that the last word in the English sentence is usually the one that carries sentence stress, which affects intonation (Birjandi and Salmani-Nodoushan 2005). It is vital to note that stress is different from intonation; it is a degree of prominence found at the level of syllables and is a factor of intonation, in addition to individual factors regarding speakers and natural ones relevant to the sound itself (Bishr 2000, p. 523). Intonation is considered as the most important phonetic characteristic which covers the whole speech. It is called secondary phonemes or suprasegmental features or prosodic features (Bishr 2000). Lexically speaking, it stands for the music of the speech, for speech production is characterized by musical colours similar to the rise and fall of musical rhythm.

Intonation in Arabic contains two types of pitch: rising and falling pitch. A falling pitch concerns the complete statements, that is, sentences with complete meaning, e.g. ***Mahmud fIlbayt*** (Mahmud is at home), and statements bearing a question and comprising a special question particle, such as ***feen*** (where), ***kayf*** (how) and ***ween*** (where), e.g. ***Mahmud feen?*** (Where is Mahmud?), and imperative statements, e.g. ***Ghadir lbayt*** (go out). In contrast, rising pitch is used in a question statement that needs a yes or no answer, incomplete statements, i.e. incomplete speech related to what comes after, e.g. ***ʔIðaʒi:t nətfahəm*** (if you come we will see). However, we may find two types of pitch in a single sentence, that is, in continuous counting numbers.

2.6.1 Functions of Intonation

Intonation comprises four linguistic functions: grammatical function, contextual semantic function, recognition function and fourth function.

- ***Grammatical function***. It is the principal factor to distinguish between syntactic patterns and different grammatical forms.
- ***Contextual semantic function***. It is used due to the different tones according to social situations.

- ***Recognition function***. It allows the speaker to recognize social and cultural strata.
- ***Fourth function***. It aims at distinguishing between meanings of the single word (Bishr 2000).

2.7 Infrequent Use of Dammah and Kasrah by Arabs

Arabs did not use the vowels, ***dammah*** and ***kasrah***, frequently in their speech due to their sensual nature which tends to use the ***fathah*** because it is very easy to pronounce. Abu Ishak argues that ***rafʕ*** of subject and ***nasb*** of object are done because they are different; thus the verb can have more than one subject but can have many objects. Consequently, the subject has the ***rafʕ*** because it is small in number compared to the object, which is abundant. So the ***fathah*** is the most frequent vowel in the Arabic language compared to the other vowels (Ibn Jini 1985, p. 48).

The ***fathah*** has no meaning but is the most used vowel by Arabs. It is easy to utter by speech organs according to grammarians, who used it a lot in their speech and quarrels as well. The investigation of Arabic language witnesses that ***fathah*** is the most frequent vowel in Arabic speech compared to other vowels. If any part of speech is put under analysis, there is no doubt that ***fathah*** is the dominant vowel. ***Surah alfatiha*** is a good demonstration of the point, where the number of occurrences of ***fathah*** exceeds the number of ***dammah*** and ***kasrah***, ***fatha*** 42 times, ***kasrah*** 31 times and ***dammah*** 4 times.

The production of short or long ***fathah*** is very easy. It does not require an effort because the air passes freely and speech organs do not provide any endeavour compared to ***dammah***, which requires rounded lips, and ***kasrah***, whose production needs the surface of the tongue to approach the hard palate (Mostapha 1992, p. 79).

2.8 Pause in Arabic

Pause is a phonological phenomenon. Lexically, it means to stop, and in speech it occurs at some place in an utterance or before uttering the next word or phrase. Contextually, it denotes ‘to cut off’ the word from what comes after by using a silence. The pause requires the completion of meaning and can be divided into categories, namely, perfect, chosen, sufficient, acceptable, good, understandable and bad avoided (Mostapha 1992, p. 79). In what follows a clear explanation is given for each type.

“The perfect pause is that whose meaning is completed and is cut off from what comes after because it is not related to it”. The sufficient pause is characterized by stopping the pronunciation but is related to the meaning; therefore it is good to stop at and start with as well. The good and understandable pause is that which is good to stop at and bad to begin with because it is related to meaning and not parsing, like the completion of a story, or information about a situation of believers in terms of expression because it is an adjective or conjoined to it, as in the following verse {(الحمد لله رب العالمين}) Praise be to Allah, Lord of the Worlds, and {الرحمن الرحيم} Most Gracious, Most Merciful, so it is very nice to make pause in these verses, because the meaning is understandable, and it is not acceptable to start with {رب العالمين} and {الرحمن الرحيم} because they are a case of genetives ***jar***; therefore it is not acceptable to start with a case of jar (Mostapha 1992).

Bad neglected pause, is characterized by a noncompletion of meaning and ambiguity in understanding such as to stop at a constructed noun without a conjoined noun, so it is not acceptable to make a pause on {لقد سمع الله قول الذين قالوا}.

2.9 Patterns of Nouns in Arabic

The formation of nouns in the Arabic morphological system consists of triliteral, quadriliteral and quintuple patterns. These nouns can be categorized into *mujarrad,* literally meaning 'bare' and *mazid,* meaning the root, to which one or two more letters are added. The triliteral is *mujarad*, like **/ʕɪnab/**(عنب), while the *mazid* is that which contains one of the ten letters, known as additional letters, and is grouped in the anagram سألتمونيهاwhich are as follows: م،ت،ن،ي،ه،اس،أل, when added to a word, the meaning changes. They can occur in the first position and contain ***(hamza)*** which is equivalent to glottal stop when pronounced and *mim* in the following words: ***/ʔusbuʕ/*** (أصبع), for 'finger', and 'maðḥab" for trend. Then the noun whose center is charged, that is, it contains a repeated letter, such as (حمص) ***hummas***, 'chickpea', then those which contain one of the aforementioned ten letters in its mid position like (طابع), which means 'stamp' and (سحاب), 'cloud' (Abu Mohammed Al Nahwi Albagdadi 1988, p. 51).

The triliteral verb is also categorized into ***mujarad*** and ***mazid almujaad*** like ***faʕala*** and ***ðahıaba***, while ***mazid*** such as ***ʔaðhıaba*** or ***rattaba*** or***ʒa:ðaba*** or ***ɪʒtaðaba*** or ***ɪnsahaba*** or ***ɪstasʕaba*** or ***takallama*** or ***ta ʒa:ðaba*** and then the quadriliteral pattern and the changes which can have due to ***huruf ziyada***, such as ***zaʕfara***, are built on the basis of four different consonants as follows:

- Four different consonants with the following symbols, 1.2.3.4, and are built from a noun like ***farqaʕa***, ***talmaða***, ***qatrana***, ***masmara*** and ***hawqala***
- Four consonants, the first resembles the third and have the following symbol 1.2.1.3 as ***tartaba***
- Four consonants where the third and the fourth (1.2.3.3) are the same, e.g. ***ʒalbaba***, ***ʃamlala*** and ***habbaba***
- Four consonants, where the first, the second and the fourth are similar (1.2.1.2) like ***zafzafa***, ***dandana*** and ***sɩarsɩara*** (Abu Mohammed 1988).

2.10 Conclusion

To wrap it up, this chapter presents an idea about the number of sounds used by Arabs, in addition to their accurate combination to attain fluency. In other words, one should apply the 12 rules called degrees of fluency stated by Al Suyuti. Furthermore the structure of the Arabic syllable is highlighted because it is vital to know, in order to apply stress which is fixed, and therefore can be governed by a rule. The compound vowel is defined afterwards and described differently from the diphthong that characterizes the English language and then is followed immediately by the infrequent use of ***dammah*** by Arabs who tend to use ***fathah***, which is very easy to produce, compared to ***dammah*** and ***kasrah***, which require an effort from the organs of speech. Finally the types of pause are presented to the reader to show their importance in the Arabic language, in addition to intonation and its function.

References

Aljahid, Abu Othman Amru Ibn Bahr died in 255H (2010). Albayan wa Al Tabyeen (the book of eloquence and oratory) (1st edn) Tahqiq Darwich Jawidi (proofread by Darwish Jawidi. Al Maktaba Al Misriya. (Egyptian Library).

Annis, I. (1997). Al Aswat alughawiyah (Speech Sounds) 4th multazamat atabaa wa anashr maktabat Anglo misriya Anglo Egyptian 165 Avenue Mohammed Farid Cairo

Arab World English Journal (2014) International Peer Reviewed Journal. 5(4). ISSN 2229-9327 AWEJ

Birjandi P, Salmani-Nodoushan MA (2005) An introduction to phonetics. Zabankadeh Publications, Tehran

Hagége, C. (1982) La structure des langues. 6eme édition collection fondée par Angoulvent encyclopédique.

Hanun M (1992) Fi Siwata Zamaniya alwaqf fi Lissaniyat alklassikiya. Dar Al Amane. Ribat (Temporal phonology, Pause in Classical Linguistics). Dar Al Amane. Rabat.

Kristeva J (1989) Language the Unknown an Initiation into Linguistics. Cambridge University Press, New York. Translated by Anne M Menke

Mostapha, I.(1992). Ihyaa Al Nahw (Revitalizing Grammar) (2 end Ed) Cairo

Quazouini, K. (1967) Talkhis Al Miftah fi Al Maani wa Al Bayan wa Al Badie (Summary of the Key in Meanings, Eloquence and Rethoric). Proofread by Yacin Souli died in 783H/1338. Egyptian Library

Radford A, Atkinson M, Britain D, Clahsen H, Spencer A (2010) Linguistics an introduction. Cambridge University Press, Cambridge

Roach P (1991) English phonetics and phonology. A practical course. Cambridge University Press, Cambridge

Roach P (1992) Introducing phonetics. Penguin, London

Sapir E (1967) *Les sens commun linguistique* Présentation de Jean-Elie Boltsanki. Les éditions de minuit. Paris

Arabic References

Ababna Y (2000) *Dirassat fi Fikhi Alugha wa Al Phonologia Al Arabiya* (Studies in philology and Arabic phonology, Ashourouk House of Publication and Distribution)

Abdelouhab H (1984) *Introduction à la phonétique orthophonique Arabe* (Collection Almoujtamaa Préface, Office des Publications, Universitaires 1 Place Centrale de Benaknoun Alger)

Abu Mohammed Ibn Sahl Ibn Saradj Al Nahwi Albagdadi died in 316 of Hijra (1988) *Al Ossoul fi anahw* (Fundamentals in grammar) proofread by Abd alhussiin Al Fatli muassasst Al Rissala (treatise institution)

Al Suyuti AJE (1987) *Al Muzhir fi Uloum Alugha Wa Anwauha.* The flowery in the Sciece of Language and its Types (Sharahahu wa dabatahu wa Anwana Mawduatihi wa Alaqa hawashih Mohammed Ahmed Al Mawla wa Ali Mohammed Albjawi wa Mohammed Abu Al Fadl Ibrahim. Proofread by Mawduatihi, Mohammed Ahmed Al Mawla, Ali Mohammed Albjawi and Mohammed Abu Al Fadl Ibrahim. Dar Al Fikr li Al Tibaa)

Almasdi A (1986) *Alissaniyat Min Khilal Al Nousous* (Linguistics from texts), 2nd edn. Al Dar Tounissia Li Anashr, Tunisian House Press

Alqamati M (1986) *Al Aswat wa Wadaifuha kouliyat Al Tarbiya, jamiat alfatih*, anashir dar alwalid Tarablus aljamahiriya aloudma (Sounds and their Functions, Faculty of Education, University of Alfatih, Published by dar alwalid Tripoli. Great Jamahiriya)

Avicenna (1983) *Rissalat Asbab Huduth alhuruf* (A Treatise on the Causes of Production of letters) Proofread by Mohammed Hacene Tayan Yahia Mir Alam. Presented and revised by Dr. Chaker Alfaham and Ahmed Rab Nafah. Dar Alfikr, Damascus

Bishr K (2000) *Ilm Al Aswat* (Phonetics) Dar Gahrib Litibaa wa anachr wa tawzii (Published by Dar Garib)

Flesh H (2007) *Alarabiya Nahw Binaa Jadeed* (Arabic Towards a New Compounding. Translated and proof read by abdSabor Shahin)

Ibn Jini. Sir Sinaat Al Irab (1985) Arabic language parsing. First part by Hacéne alhindawi akhdar alkqlam. Damascus

Kaddour AM (1999) *Mabadie alissaniyat* (Principles of linguistics), 2nd edn. Dar alfikr Almoaasir, Beirut, Lebanon

Saaran M (1994) *Ilm Alugha muqadimah li Alquarie Alarabi* (Science of Language an Introduction to the Arabic Reader). University of Alepo

Touiji M (2001) *Almuajam Al Moufassal fi Ouloumi Alugha*. Isnad raji alasmar wa murajaat Emil yakob (Detailed Dictionary of Linguistics proof read by Emil Yacok, 1st edn. Beirut, Lebanon, Dar alkutub alilmiya)

Zamakhshari (2001) *Sharh Al Mufassal* (Al Mufassal explanation. Proofread by Emil Badie Yakob, 1st edn. The Fifth section. Manshourat Mohammed Ali Baydon Linashr kutub Aljamaa wa Al sunna. Dar alkutub Al Ilmiya, Beirut, Lebanon)

Chapter 3
The English Phonological System

3.1 Introduction

Like many linguists, cognitive linguists study language for its own sake; therefore they try to describe and explain its systematicity, its structure, the different functions it serves and its realization. In this respect, cognitive linguists in studying language attempt to investigate the patterns of conceptualization, i.e. language offers a window into cognitive function scrutinizing the nature of the structure and organization of thoughts.

The most important thing to mention is that cognitive linguistics differs from other approaches to the study of language, in that, it reflects certain fundamental properties and design features of the human mind (Evans and Green 2006, p. 6).

3.2 Aim of Language

We rely upon language during our whole life to perform a set of functions. So, how we would accomplish all these functions without language even in one single day? For instance, to buy things from a shop, asking for information, sending a message, expressing ideas, making agreement and disagreement, expressing happiness and displeasure, how to use the Internet and mobile phones, etc. (Evans and Green 2006).

In all these situations, language is the only solution which allows a very rapid and effective expression; thus it provides developed means for encoding and transmitting complex ideas. In fact the notion of encoding and transmitting is very important in language, because they are associated with the symbolic and interactive functions of the language (Evans and Green 2006).

M. Dib, *Automatic Speech Recognition of Arabic Phonemes with Neural Networks*, SpringerBriefs in Applied Sciences and Technology,
https://doi.org/10.1007/978-3-319-97710-2_3

3.2.1 Language as a Symbol

The most difficult function of language is the expression of thoughts, i.e. it encodes and puts our thoughts into action, so it uses symbols which can either be segments of words like **(dis)** in the word **(distaste)** or complete words, such as **cat**, **run**, **tomorrow**, or sentences such as *He couldn't write a pop jingle, let alone a whole musical.*

Symbols refer to an association between form and meaning; in other words, they are represented in spoken, written or signed forms and carry conventional meanings. The form can be a sound represented by the International Phonetic Association or a grapheme represented orthographically or a sign of a sign language (Evans and Green 2006).

3.2.2 Language as Interaction

Language performs an interactive function in our daily life, and as mentioned before, it is an association between form and meaning, but this is not sufficient because it should be recognized and understood by others in the community.

A language is first and foremost used for communication and therefore requires transmission of an idea from a speaker, to be decoded and interpreted by the listener, processes that require a construction of rich conceptualizations (Evans and Green 2006).

3.3 English Sounds

English contains 44 phonemes, 20 vowels and 24 consonants. Vowels are categorized into three types: short vowels, long vowels and diphthongs. The number of short vowels is seven, and the number of long vowels is five, but the number of diphthongs is eight.

3.3.1 Vowels

The production of speech requires three systems for the airstream to pass through, respiratory, phonatory and articulatory systems, where four main resonators are found, pharyngeal cavity, oral cavity, labial cavity and nasal cavity. The resonators are considered as passages through which speech sounds move and can either be obstructed or freed (Birjandi and Salmani-Nodoushan 2005, p. 55).

The presence or absence of obstacle airstream can alter the speech sound into vowels, glides (semivowels), consonants or liquids (Birjandi and Salmani-Nodoushan 2005). In what follows vowels are presented in detail.

Vowels are very important in all languages of the world because they are necessary to form syllables alone (Ashby and Maidment 2005, p. 70), e.g. eye, awe, can owe and or; however consonants cannot form words alone (Peinado and Segura 2006, p. 134).

3.3.1.1 Characteristics of Vowels

The sound is not called a vowel unless there are three criteria: degree of openness of the oral cavity, degree of laxity of vocal tract muscles and amount of duration of articulation. If these characteristics are not found in a sound, then it is a consonant and not a vowel (Birjandi and Salmani-Nodoushan 2005).

3.3.1.2 Types of Vowels

English vowels can be classified into two categories: long vowels and short vowels. Long vowels differ from short vowels in terms of length, that is, their production requires a time duration: thus their representation is done schematically by putting a colon (:) immediately after them. RP English has five long vowels and seven short vowels. Tables 3.1, 3.2 and 3.3 display these types of vowels (Birjandi and Salmani-Nodoushan 2005) (Tables 3.1, 3.2 and 3.3).

3.3.1.3 Schwa

It is the most frequently occurring vowel in English speech, and it always occurs in unstressed syllable, it is a mid-vowel, and it is described as being one unrounded central vowel, i.e. (between front and back), and mid, that is, between close and open (Roach 1992, p. 96). It does not require any energy in its pronunciation. As for

Table 3.1 Short vowels

Vowel	Word	Phonetic transcription
/ɪ/	Hit	/hɪt/
/e/	Bed	/bed/
/æ/	Cat	/kæt/
/ə/	Ago	/əgəʊ/
/ʊ/	Book	/bʊk/
/ʌ/	Cut	/kʌt/
/ɒ/	Hot	/hɒt/

Table 3.2 Long vowels

Vowel	Word	Phonetic transcription
/iː/	Sheep	/ʃiːp/
/ɜː/	Sir	/sɜː/
/ɔː/	More	/mɔː/
/ɑː/	Far	/fɑː/
/uː/	Boot	/buːt/

Table 3.3 Description of vowels according to the vowel chart

Vowel	Description	Vowel	Description
/iː/	High front vowel	**/uː/**	High back vowel
/ɪ/	High-mid-front vowel	**/ʊ/**	High-mid back vowel
/e/	Mid-front vowel	**/ɔː/**	Mid-back vowel
/æ/	Low front vowel	**/ɑː/**	Low-back vowel
/ɜː/	Mid-low mid-vowel	**/ɒ/**	Mid-low-back vowel
/ə/	Mid-mid vowel	**/ʌ/**	Low-mid vowel

its distribution in the word, it can occur in initial, medial and final position, e.g. about/əbaut/, accurate/ækjərət/ and character /kærəktə/ (Roach 1991, p. 64).

Unlike many languages, where all vowels can be distributed in both stressed and unstressed syllable, English, however, distinguishes between both types of syllable, that is, the vowel schwa can occur only in unstressed syllables (Birjandi and Salmani-Nodoushan 2005) (Tables 3.4 and 3.5).

3.3.2 *Consonants*

English has 24 consonants classified in terms of place of articulation, manner of articulation and voicing. The [b] sound, for example, is a bilabial voiced plosive, whereas [s] is a voiced alveolar fricative consonant (Collins and Mees 2003, p. 35).

3.3.2.1 Classification of Consonants

Consonants are classified in terms of energy of articulation, place of articulation and manner of articulation, for example, /b/ can be described as a lenis bilabial plosive, /s/ as a fortis alveolar fricative and /ŋ/ as a velar nasal. Before embarking on the discussion of place and manner of articulation, it is worth mentioning that there are two types of articulators which can produce sounds (Collins and Mees 2003).

Table 3.4 Location of schwa in syllables

Word	End of word	Phonetic trans (energy)	Phonetic trans (less energy)	Word category
Intimate	ate	/ıntımeıt/	/ıntımət/	Adjective
Accurate	ate	/ækjəreıt/	/ækjərət/	Adjective
Desolate	ate	/desəleı/	/desələt/	Adjective
Private	ate	/Praıvıt/		Adjective (exception)
Potato	o	/pɒteıtəu/	/pəteıtəu/	Noun
Carrot	o	/kærɒt/	/kærət/	Noun
Ambassador	or	/æmbæsədɔ/	/æmbæsədə/	Noun
Forget	or	/fɔ:get/	/fəget/	Verb
Opportunity	or	/ɒpɔ:tjunıtı/	/ɒpətjunıtı/	Noun
Settlement	e	/setlment/	/setlmənt/	Noun
Violet	e	/vaılet/	/vaılət/	Adjective
Postmen	e	/pəustmen/	/pəustmən/	Noun
Strongest	er	/strɒŋgɜ:/	/strɒŋgə/	Superlative
Perhaps	er	/pɜ:hæps/	/pəhæps/	Adverb
Support	u	/s^pɔ:t/	/səpɔ:t/	Noun
Autumn	u	/ɔ:t^m/	/ɔ:təm/	Noun
Halibut	u	/hælıbət/	/hælıb^t/	Noun
Thorough	ouh	/Θ^ruə/	/Θ^rə/	Adjective
Borough	ouh	/b^rəu/	/b^rə/	Verb
Gracious	ious	/greıʃə/	/greıʃə/	Adjective
Callous	ous	/kæləs/	/kæləs/	Adjective

Table 3.5 Diphthongs

Diphthongs in ə ending			Diphthongs in ʊ ending			Diphthongs in ɪ ending		
Word	Ph transc	Diph	Word	Ph trans	Diph	Word	Ph transc	Diph
Poor	/pu ə/	uə	out	/aʊt/	aʊ	play	/ple ɪ/	eɪ
Bear	/beə/	eə	owe	/ʊə/	ʊə	boy	/bɔɪ/	ɪɔ
Here	/hɪə/	ɪə				try	/tra ɪ/	ɪa
					Total number	08		

3.3.2.2 Active and Passive Articulators

There are two types of articulators: active and passive. Active articulators are organs that move in the articulation, but passive articulators are the target of articulation (Collins and Mees 2003).

Some sounds require contact between the two articulators, such as [t] and [k], and other sounds are made without contact, such as s and x and r, but for h, b, p and

Table 3.6 Manner of articulation

Manner	Phonemes	Reminding sentence
Plosives	P,t,k,b,d,g.	Please take bright dagger
Fricatives	ʃ, f, z, v, h, s, ð ʒ, Θ.	**Sh**e **f**eels **v**ery **h**appy **s**eeing **th**em measuring leng**th**
Affricates	ʧ,ʤ	Charge
Nasals	m,n,ŋ	Meaning
Approximant	j r w l	you are Wili
Lateral	L	
Glide	w j	
Rotic	R	

Table 3.7 Place of articulation

Place of articulation	Phonemes	Reminding sentence
Bilabial	p,m,b,w	**P**ut on **m**y **b**lack **w**ear
Labiodentals	f,v	**F**ill **v**acuum
Interdental	Θ,ð	**Th**ank **th**em
Alveolar	t,d, s z, n r, l.	**T**ommy **d**efend**s** **z**one **n**on **l**uc**r**atively
Palate alveolar	ʃ,ʧ,ʤ,ʒ.	**Sh**e **ch**ange**s** **g**enre
Palatal	J	**Y**es
Velar	k, g ŋ	**G**o **K**i**ng**
Glottal	h	**H**i

m, it is very hard to distinguish between both types of articulators, because [h] is laryngeal and [b], [p] and [m] are bilabial; the two lips are pressed together (Collins and Mees 2003).

Place of articulation. The place of articulation is considered as an essential part to describe consonants, for it informs us about the direction towards the sound is directed. There are nine places of articulation in the English language: lips, lips and teeth, teeth, alveolar ridge, alveolar ridge and palate, soft palate, hard palate, larynx and lips and soft palate. Sounds produced in these places are called bilabial, labio-dental, interdental, alveolar, palato-alveolar, velar, palatal, glottal and labiovelar.

Manner of articulation. There are six degrees of obstruction, so English consonants are described in terms of six manners of articulation which are plosive, fricatives, affricates, nasals, approximant and lateral.

The table below clearly shows manners of articulation with some reminding sentences that help keep all the types of consonants in mind (Tables 3.6 and 3.7).

3.3.3 Syllable

The syllable has no suitable explanation, yet to try to elucidate it, one should observe that it is a clear peak surrounded by a group of consonants.

But sometimes boundaries of syllable are neglected, and the idea to not consider some peaks, such as /s/ in stop and /k/ in cry as syllables, is to be bypassed (Duanmu 2008, p. 36).

There is another definition which argues that the syllable is associated with chest pulse but ignores syllable boundaries. Gimson, on the other side, rejects the pulse theory and claims that the double chest pulse is not clearly visible in the word *seeing* [**siːɪn**]and the pulse theory is unable to determine whether the word *beer* [**bɪər**] is made up of two syllables in American pronunciation. This controversy provokes some hesitation about the nature of the syllable, i.e. whether syllables are linguistic units or not. Chomsky and Halle, Steriade (1999), Gimson (1970) and Belvins (2003), as cited in Duanmu (2008), reject the idea that syllables are phonological units (Collins and Mees 2003).

Notwithstanding, syllable may be noticeable in some instances; there is a consensus among people that the word Canada contains three syllables and the word America entails four syllables (Duanmu 2008, p. 36).

3.3.4 Stress

Stress in English is used in words that contain one or more than one syllable; it is called free stress, i.e. it can be governed by a rule, because it can fall on the first syllable, second, third or fourth. The characteristics found in unstressed syllables are different from those in stressed syllables; in other words stressed syllables are produced with a strong effort, and air is pushed strongly by chest muscles, compared to unstressed syllables.

English is a stress-timed language, compared to syllable-timed languages, such as French and Italian, where syllables have the same proportion of time in terms of stress. Speakers of syllable-timed languages tend to not understand the reason for using rapid speech from English speakers in a particular sentence because syllables in syllabic languages have the same duration of time in terms of stress. However in English stress falls on particular words only, for instance, the modal verb (can) does not bear stress in the affirmative form (Birjandi and Nodoushan).

Stress is considered an important feature in any language, because it helps in displaying contrast; therefore it reduces ambiguity. In careful speech, for instance, the pause is not detectable between sentences and words; therefore it seems very long to the listener who is not familiar with the language, so stress in this case can help to display the boundaries of words and sentences.

In French and even Persian, stress usually falls on the first syllable, whereas in other languages, such as Hungarian and Icelandic, stress falls on the last syllables, so stress can help in understanding languages in the sense that it can display the meaning and the grammatical category of the word. The word convert is a good demonstration of the point. If it is stressed on the first syllable, then it is a noun, but if it is stressed on the second, then it is a verb; Table 3.8 displays the function of stress.

Stress in isolated words is called word stress, and not all words carry stress; some words carry strong stress, others carry weak stress and others do not have stress at all.

Table 3.8 Function of stress

Word	Verb	Noun
Convert	/kən'vɜ:t/	/'kɒnvət/
Contrast	/kən'træst/	/'kɒntræst/

As for the words that carry stress, they are lexical words, such as adjectives, verbs and nouns; in contrast, words that do not carry stress are definite articles, such as *the* and *a*, connectors like *and* and *but*, pronouns such as *me* and *them*, prepositions like *from* and *with* and auxiliary verbs, such as *do*, *be* and *can*. Such words bear less information though they relate sentences; however lexical words are considered as fundamental for bringing information. It should be pointed out that there are only two types of grammatical words that carry stress: Wh question, e.g. *where*, *why*, *how*, and *which*, and demonstratives, e.g. *this*, *that*, and *there* (Collins and Mees 2003, p. 19).

3.3.5 Intonation

It is the use of pitch to transmit linguistic information, feelings and attitudes, and it is used to express two things:

- Variation of pitch to transmit or alter a meaning
- Synonym of prosody which stands for sound quality, rhythm and rise of the sound (Roach 1992, p. 56)

- Intonation refers to when, why and how the speaker rises or keeps his pitch once speaking (Birjandi and Salmani-Nodoushan 2005).

3.3.5.1 Types of Intonation

Roach (1992) states that there are five types of intonation: rising, falling, level, rise-fall and fall-rise. However, other phoneticians add another type called takeoff intonation (Birjandi and Salmani-Nodoushan 2005).

Rising intonation. It is characterized by a clear rise of pitch after the stressed syllable in the final word located in a sentence, and it is used in yes/no questions, statement questions and tag questions (Birjandi and Salmani-Nodoushan 2005).

Falling intonation. It is characterized by a clear fall of pitch after the stressed syllable in the final word found in a sentence; simply put, it is used in many cases such as Wh questions, confirmatory tag questions and statements; sometimes these statements are imperative.

Level intonation. It is one type of intonation in which the pitch is constant, that is, not modulated, as opposed to rise-fall or fall-rise tone. Additionally, level intonation is used in a very strict usage, such as frozen style used by the bishop who reads the holy book to church-goers and sometimes in the courts of law (Birjandi and Salmani-Nodoushan 2005); this is as far as statements are concerned. But in terms of one syllable utterance, level tone is used to answer some boring questions, mainly that of insurance service or when the teacher calls his pupils from a register, i.e. in a routine situation (Roach 1991).

Fall-rise intonation. It is characterized by a combination of the fall and rise of pitch, and it is used in soothing and politeness, that is, parents use it to calm their children when they are in an uneasy way. In addition people use it for polite answers, e.g. thank you (Birjandi and Salmani-Nodoushan 2005).

Rise-fall intonation. This type of intonation is used in listing and calming children, e.g. don't cry and I'll take you to the park.

Takeoff intonation. This kind of intonation can be compared to the way the airplane runs before takeoff; in other words, it is characterized by the takeoff pitch after the most significant contrastive stress of the sentence, i.e. the speaker moves from a regular or level tone and gradually raises the pitch. The full rising pattern of the pitch is related to the choice of the speaker. Additionally, this type of intonation is mostly used with negatively charged emotion such as grumbling, e.g. ***you shouldn't give him all that money you silly boy***. Cursing and blasphemy also use takeoff intonation because they are a sort of negative emotional expression (Birjandi and Salmani-Nodoushan 2005).

3.3.5.2 Functions of Intonation

Intonation has four linguistic functions which are as follows: focusing, grammatical, function and discourse (Collins and Mees 2003).

Focusing function: It permits the speaker to emphasize the significant information in the sentence (Collins and Mees 2003).

Grammatical function: It is the function which permits to the speaker to distinguish the syntactic relationship, e.g. phrases and clause boundaries, Wh questions and statements (Collins and Mees 2003).

Attitudinal function: It allows the speaker to express an attitude, such as feeling, emotion, finality, confidence, surprise, etc., in other words, something which has no relation with semantic content, on top of what is being said (Collins and Mees 2003).

Discourse function: Encompassing diverse matters; the organization of conversations between two or more speakers, e.g. a signal for turn-taking (Collins and Mees 2003).

3.4 Conclusion

This chapter focuses on the English phonological system, the reason being, of course, that this research deals with a contrastive study in English and Arabic at the phonological level. The chapter starts with a brief introduction where language is presented as being associated with two functions, i.e. a symbolic function and interaction, then the English sounds (vowels and consonants) are highlighted, and then syllable, stress, intonation and functions of intonation are mirrored.

References

Ashby M, Maidment J (2005) Introducing Phonetics Science (1st Ed). Cambridge University Press
Birjandi P, Salmani-Nodoushan MA (2005) An introduction to phonetics. Zabankadeh Publications, Tehran
Collins B, Mees IM (2003) The phonetics of English and Dutch, 5th edn. Brill, Leiden
Duanmu S (2008) Syllable structure, the limits of variation. 1st edn. Oxford University Press, Oxford
Evans V, Green M (2006) Cognitive linguistics an introduction. Edinburgh University Press, Edinburgh
Peinado MA, Segura JC (2006) Speech recognition over digital channels robustness and standards. Wiley, Chichester
Roach P (1991) English phonetics and phonology. A practical course. Cambridge University Press, Cambridge
Roach P (1992) Introducing phonetics. Penguin, Harmondsworth

Chapter 4
A Contrastive Phonological Study of English and Arabic

4.1 Introduction

Languages are different from each other in terms of systems: phonetic, phonological, morphological, syntactic and semantic. This is due to the physical and the social environment in the Sapirian theory and to the parametric variations in Chomskian theory, where speakers use nouns and verbs, for instance, as heads or tails in sentences or the opposite. This denotes that the speakers' brains are in a certain state, in the sense that language is a cognitive system which is a part of any normal human beings mental or psychological structure. The focus in this chapter is made on the different structure of the speakers' brains with regard to the diphthong. For example what is considered as a diphthong, i.e. a vowel, in English is regarded as a syllable in Arabic.

Schwa is one vowel which is the most frequent in English speech, and it is difficult to understand by learners of English, mainly Algerian students who find it difficult to locate its position.

To solve this problem, a contrastive study in English and Arabic is done to examine schwa in the final position and to assign its equivalent in Arabic, so as to help students detect it in words and pronounce it correctly. To this account, phonemic taxonomy is introduced, and then a practical survey relying upon a contrastive study in similar words taken from English and Arabic is applied afterwards. In what follows Harris taxonomy is presented.

M. Dib, *Automatic Speech Recognition of Arabic Phonemes with Neural Networks*, SpringerBriefs in Applied Sciences and Technology,
https://doi.org/10.1007/978-3-319-97710-2_4

4.2 Phonemic Analysis (Taxonomic Phonology)

4.2.1 Definition

Phonemic analysis or taxonomic phonology stands for the inventory of phonemes including the rules of realizing the phones representing them and the distribution that characterizes the syntactic structure of a morpheme. In other words, three things should be applied: phoneme inventory that concerns every language; allophonic rules, i.e. a set of rules that concerns each phoneme; and the acceptable sequence of phonemes called phonotactics (Lass 1984, p. 21).

4.2.2 Taxonomic Phonemics

Doubtless, the primary object in a study of structural linguistics is the sound pattern which has been studied either in complete or relative isolation from syntax. Taxonomic phonemics hinges upon segmentation and classification of the linguistic elements within the speech. It entails four principles: linearity, biuniqueness, invariance and local determinacy (Chomsky 1970, p. 75).

Linearity: Speech is a sequence of phonemes whose features are distinctive and redundant. Consequently the phonetic value of phones is the same as the sequence of phonetic value of these phonemes. There are two things in linearity: that there is a correspondence between each phoneme and a phone or a sequence of consecutive phones and that the same linear order of the corresponding phones is presented in the linear order of phonemes in any particular utterance (Chomsky 1970).

Biuniqueness: It puts its emphasis on the strong correspondence between the phone and the phoneme, that is, each sound corresponds to a symbol and vice versa (Chomsky 1970).

Invariance: Phonetic invariance applies to the initial segmentation of utterances. Once an utterance is segmented according to its differences with other contrasting utterances, the resulting segments can be recognized in other utterances by their phonetic attributes (Chomsky 1970).

Local determinacy: It means a local one to one correspondence, that is, the phonemic representation of a particular phonetic form is done by a purely phonetic consideration or by some neighbouring sounds (Chomsky 1970).

4.3 Sounds

To make contrast between Arabic and English, words require sequence to the number of sounds in both languages. Arabic contains 28 consonants, though the number of letters is 29. The reason is that the alif is a symbol used to represent the long vowel, but not a consonant. The consonants start from the glottal stop referred to as ***hamza*** and end with the /j/ sound. Contrariwise, English has 24 consonants and 20 vowels. Tables 4.1 and 4.2 display the Arabic and the English sounds.

Table 4.1 Arabic sounds

Phonetic symbol	Equivalence	Written form
ʔ	**HAMZAH**	ء
b	**BA**	ب
t	**TA**	ت
θ	**THA**	ث
ʒ	**JIM**	ج
h	**HA**	ح
x	**KHA**	خ
d	**DAL**	د
ð	**DHAL**	ذ
r	**RA**	ر
z	**ZAY**	ز
s	**SIN**	س
ʃ	**SHĪN**	ش
s̞	**SAD**	ص
d̜	**DHAD**	ض
t̜	**THA**	ط
ð	**ZHĀ**	ظ
ʕ	**AI N**	ع
ɤ	**GHAIN**	غ
f	**FA**	ف
q	**QAF**	ق
k	**KAF**	ك
l	**LAM**	ل
m	**MIM**	م
n	**NUN**	ن
h̜	**HAMZAH HĀ**	ه
w	**WAW**	و
j	**YA**	ي

Table 4.2 English sounds

Consonants	Vowels
b	**Short vowels**
d	ʌ
f	æ
g	e
h	ə
j	ɪ
k	ɒ
l	ʊ
m	**Long vowels**
n	ɑ:
ŋ	ɔ:
p	ɜ:
r	u:
s	i:
ʃ	**Diphthongs**
t	aɪ
ʃt	aʊ
θ	oʊ
ð	eə
v	eɪ
w	ɪə
z	ɔɪ
ʒ	ʊə
dʒ	

4.4 English Corpus

	Video—The man with the bionic eye
Film maker	BBC Inside Out London
Date	March 2009
Subject:	How science fiction is becoming a reality with a remarkable medical breakthrough for the blind

4.4.1 Bionic Eye

A London eye hospital is at the forefront of a unique trial that has the potential to restore blind people's sight. BBC Inside Out was given exclusive access to groundbreaking work taking place at Moorfields Eye Hospital.

Most of us take our sight for granted, but two million people in the United Kingdom have some sort of problem with their sight.

For 25,000 people in London, blindness and the condition retinitis pigmentosa are a daily reality.

At Moorfields Eye Hospital, surgeons are pioneering a unique invention that enables surgeons to fit patients who have lost their sight with a bionic eye.

4.4.2 Total Darkness

Ron is one of the capital's 25,000 blind residents, and because of a hereditary condition, he has been living in total darkness for the last 30 years.

He is now one of just three people in the country to have been fitted with a revolutionary bionic eye, which is having a dramatic impact on his life.

Thanks to this implant, he is now able to see different shades of light. He can now walk along a white line painted on the ground and even sort out his socks into white, black and grey piles.

4.4.3 Lighting Up Ron's World of Darkness

The bionic eye works by capturing light onto a video camera in the patient's glasses, which sends a wireless signal to the implant which stimulates the optic nerve.

This medical breakthrough has the potential to not just radically transform the lives of the blind but also to improve the normal sight powers of future generations.

This is no overnight miracle cure to blindness, and at the moment it's only being trialled on a very specific group.

What is so exciting about the bionic eye is its potential for the future.

"Gregoire Cosendai" from Second Sight says: 'In 50 years' time I hope that people will be able to read with this system and it's not unthinkable that in the distant future people will have a retina implant that can provide them with better vision than normal seeing people.

BBC Inside Out asked patient Ron about his experiences with the 'bionic eye'.

'Bionic eye'	
	Video—Bionic eye
Filmmaker:	BBC Inside Out London
Date:	March 2009
Subject:	Meet the man with the 'bionic eye' who is giving visually impaired Londoners hope for the future

Q and A interview with Ron

Presenter Matthew Wright and Ron:

How are you getting along?

Ron: Slowly but surely. They said 'let there be light', and there was light. For 30 years I've seen absolutely nothing at all, it's all been black, but now light is coming through.

It is truly amazing. They're wonderful people these scientists.

It's exciting. And after you've seen nothing for 30 years but darkness, suddenly to be able to see light again is truly wonderful.

It's like the future coming to us now in the present, isn't it?

Ron: It is. My one ambition at the moment is to see the Moon, to go out on a nice clear evening and to be able to pick up the Moon.

Whether I'll be able to do it or not, I don't know, but I'm relying on these scientists.

How did you lose your sight, Ron?

Ron: It's a family thing, it's one of these hereditary complaints, called Retinitis Pigmentosa, normally known as tunnel vision.

And, basically, your peripheral vision starts to disappear until you're left with central vision, which means you can recognise someone 50 yards down the road and wave to them and walk in to a lamppost which is only six inches at your side.

And then, eventually, my central vision went, and I was registered blind in 1979. I got a guide-dog in 1980, and I've never looked back from that.

In terms of your sight now, you have no sight at all?

Ron: None whatsoever - everything in black.

The reason I think that it took me a long while to make up my mind whether I wanted to go for this experiment, because it meant a three or four hour operation.

And you know, obviously you were in hospital and the scientists didn't know exactly what the results were going to be, and you didn't...

How did you first hear about what was going on at Moorfields?

Ron: Basically, through my wife. She used to work for the Guide-dog Association, and she keeps up to date with Retinitis Pigmentosa Association, and we get a magazine every quarter.

They mentioned that there was going to be a seminar held for this advanced technique by Second Sight. So she took the details and eventually she persuaded me that I should at least go along and listen to what was being said... Before I knew where I was, she'd put my name on the list.

Once your wife put your name forward, what was the process?

Ron: You were interviewed and... you had to comply with five criteria... I think you had to be completely blind, you had to live within two hours of Moorfields. Obviously you had to be able to convey to the scientists what you could see, and your ganglions had to be in order.

Did you have concerns about an operation?

Ron: No, if I'm absolutely honest with you. It's always been my idea that dying can be painful and the one way you can eliminate a painful death is by going under an operation, because you know nothing about it...

What did the operation actually involve, Ron?

Ron: It's the right eye they operate on. They open the eye and they implant a small - a ray, and they tack it to the back of the retina and it contains 60 tiny electrodes.

Each of those electrodes is connected to a wire, and that wire is brought out from the side of the eye, below the cheekbone, where you can't see it.

'Let there be light': eye operation.

A little radio receiver, for want of a better word, is placed there, and a piece of donated sclera, the white of an eye, is used as a sort of belted across the eye to hold it in place.

Then you come across the glasses which contain a little camera in the nose piece, and it also has a radio link.

They call it a RF link, which is attached to the cheek, which is on the glasses but pressed against the cheek, and a cable runs from the camera to a small computer, which you can wear on a belt, no bigger than a packet of cigarettes.

Then the information that the computer receives is fed up to the induction coil, as they call it - the size of a 50 pence piece on the cheekbone.

The radio signals are transmitted to the link on the outside of the eye.

Then it's by the cable to the 60 electrodes at the back of the eye, which when they're agitated or lit up make the retina respond so you can actually pick up light.

Making Progress—Ron's New Bionic Eye

Any regrets?

Ron: No. It's a great privilege and an honour to be able to take part in an experiment such as this, hoping that the outcome is going to be able to bring sight to people, like myself, that are completely blind.

What do you see, what can you see?

Ron: The one advantage it has at the moment is more on my wife's side, because I can now sort the washing out. It gives me grades of bright light to black, and anything in between.

I can actually sort out white socks, grey socks and black socks, but as far as the washing is concerned it's just a question of things are either white or they're coloured, and that suits my wife down to the ground.

Can you see shapes?

Ron: No - there is no shapes as far as I'm concerned... You can't see print or anything like that.

Vision - as you know it – isn't there. I can pick up a window, I can pick up possibly a door frame, but as for - as for useful vision to enable me to move around comfortably, I'll stick to the guide-dog.

But this is early days, it's only 6 months.

Will this improve in the future?

Ron: I sincerely hope so. I think it's really working with the scientists and also educating your brain to understand what you're seeing.

At the moment I need someone to tell me what I'm looking at... but I'm hoping that my brain will begin to put the pictures together to enable me to understand what I'm looking at.

What do you see, what can you see?

Ron: The one advantage it has at the moment is more on my wife's side, because I can now sort the washing out. It gives me grades of bright light to black, and anything in between.

I can actually sort out white socks, grey socks and black socks, but as far as the washing is concerned it's just a question of things are either white or they're coloured, and that suits my wife down to the ground.

4.5 Arabic Corpus

سرطان الكلى.. أساليب علاجية جديدة

التاريخ: 20/3/1431 الموافق 06-03-2010 | الزيارات: 901. علاجات فعالة تمنح المرضى أملا كبيرا في حياة أطول المختصر / يعتبر سرطان الكلى من المشكلات الصحية القاتلة، فهو المسبب الرئيسي الثاني للوفاة في العالم. وتشير الإحصائيات الخاصة بسرطان الجهاز البولي والتناسلي إلى أن هناك زيادة سنوية في مرضى سرطان الكلى في الوطن العربي مواكبة للزيادة الحاصلة في العالم. لقد وجد أن حالات سرطان الكلى قد زادت بنسبة 52 في المائة خلال الفترة بين عامي 1983 و2002، أي من 7.1 إلى 10.8 حالة لكل 100 ألف شخص. وارتفعت معدلات الوفيات أيضا، وبشكل خاص بين أولئك المصابين بالأورام الأكبر حجما من 7 سنتيمترات، فارتفعت من 1.2 إلى 3.2 لكل 100 ألف شخص. كما تشير كثير من الإحصاءات العربية إلى زيادة نسبة مرضى الكلى في الوطن العربي، فمثلا تبين إحصاءات السجل الوطني السعودي للأورام أن معدل الإصابة بسرطان الكلى في المملكة العربية السعودية وصل إلى 200 حالة سنويا، وأكثرها من مكة المكرمة والرياض والمنطقة الشرقية. * مؤتمر طبي * وللوقوف على أهم المستجدات العلمية والدوائية لسرطان الكلى، أقامت الجمعية السعودية لجراحة المسالك البولية بالتعاون مع الجمعية السعودية للأورام مؤتمرا طبيا في مدينة شرم الشيخ في الفترة من 10 إلى 12 فبراير (شباط) الحالي. وضمن حديثه لـ«صحتك»، أوضح رئيس المؤتمر أ.د. أشرف أبو سمرة، استشاري جراحة

أورام المسالك البولية - مدينة الملك عبد العزيز الطبية في جدة، أن الهدف الأساسي للمؤتمر هو رفع مستوى الوعي لدى الأطباء وبالتبعية لدى المرضى،
والاهتمام بترسيخ قاعدة «الوقاية خير من العلاج» والإشارة إلى أهمية وخطورة مرض سرطان الكلى، وخصوصا بعد زيادة نسبة الإصابة بالمرض في العالم
أجمع. وأكد د. أبو سكرة على الأشخاص الذين لديهم تاريخ مرضي عائلي لسرطان الكلى أن يخبروا أطباءهم بذلك ويقوموا بعمل الاختبارات اللازمة بصورة
متكررة، فالقيام بدور إيجابي نحو المحافظة على الصحة يزيد من فرص اكتشاف المرض مبكرا. * علاجات فعالة * يضيف الدكتور أبو سكرة أن هناك
ويمكن إعطاء أكثر من نوع من العلاجات في الوقت نفسه .Renal Cell Carcinoma خيارات كثيرة متاحة لعلاج مرض سرطان خلايا الكلى
اعتمادا على المرحلة المرضية للسرطان. فهناك جراحة الاستئصال الجذري للكلية، استئصال الكلية بالمنظار، إزالة الثانويات المنتشرة، العلاج الإشعاعي، العلاج
وهذا الأخير يعد بارقة أمل لهؤلاء المرضى، حيث أثبت فعالته في علاج .(Target Therapy) الحيوي، والعلاج بالأدوية المستهدفة بالرحمة
والأدوية الأخرى المماثلة من النوع نفسه. * خيارات علاجية متعددة Nexavar (sorafenib) هذا النوع من السرطانات. مثل عقار نيكسافار
* الجراحة هي الخط الأول في علاج سرطان خلايا الكلى مع كثير من المرضى، ولها احتمالية الشفاء، إلا أن السرطان متوسط أو عالي الخطورة غالبا ما
يعود بعد الجراحة (في 35 - 65 في المائة من الحالات). وهذه الحقيقة محددة بالدراسة التالية، التي شملت 1671 مريضا لديهم إكلينيكيا سرطان خلايا
(Leibovich BC et al. الكلى من نوع «الخلية الواضحة» موضعي، وفي جانب واحد من الجسم، خضعوا لجراحة استئصال جذري للكلى
Cancer 2003: 97:1663 - 71). وقد حدث انتشار للمرض في الجسم في 479 مريضا في خلال 1.3 سنة في المتوسط. وكان
متوسط فترة البقاء بدون انتشار للورم 86.9 في المائة عند سنة واحدة، و77.8 في المائة عند 3 سنوات، و67.1 في المائة عند 10 سنوات.
أمل جديد يقول البروفسور بيتر مولدرز من «مركز نيميجين الطبي» بجامعة رادبوود في هولندا إن «المجموعات المنتقاة من المرضى لديها احتمال ضعيف
لتطور المرض بعد استئصال الكلى، وإن العلاج المساعد (بعد الجراحة)، بواسطة العلاج الإشعاعي، والعلاج الهرموني، والعلاج الكيميائي المعهود،
والسيتوكينات أظهر كفاءة ضعيفة في علاج سرطان خلايا الكلى. وعلى الرغم من ذلك فهناك أمل جديد»! ويضيف: «إن الأدوية الموجهة مثل مثبطات
عن طريق الفم جيدة التحمل، وفعالة في علاج سرطان خلايا الكلى، وتناسب العلاج طويل Tyrosine Kinase Inhibitors تيروزين كيناز
الأمد. هذه الأدوية يمكنها بالتالي أن توفر خيارات للعلاج في المستقبل بعد الجراحة». والمرحلة التالية لكثير من التجارب السريرية مع هذه الأدوية هي قيد
كعلاج (Sunitinib) أوسنيتينيب (sorafenib) وتعني: استخدام سورافينيب(ASSURE) التنفيذ حاليا، بما في ذلك التجربة أشور
placebo وتعني: مقارنة سورافينيب مع العقار الوهمي) SORCE مساعد في حالات سرطان الكلى ذات النتيجة غير المواتية)، والتجربة سورس
في المرضى الذين يعانون من سرطان الخلايا الكلوية الأولي بعد استئصاله مباشرة). وأشار د. مولدرز إلى أنه ما زالت هناك حاجة إكلينيكية واضحة للعلاج
المساعد. * الحل الأمثل * يقول البروفسور يورجين جيشوند، من «مركز ريكتس در إيزار الطبي» بجامعة ميونيخ الفنية بألمانيا إن هناك عددا من
الاستراتيجيات المتاحة حاليا وتحت التطوير من أجل تحسين نتائج العلاج. ويشمل ذلك: تحديد الجرعة الأمثل، علاج الأعراض الجانبية، والمزج والتعاقب
الأمثل للعلاجات. ويضيف: أن استخدام مزيج من الأدوية الموجهة يمكن أن يحسن النشاط الإكلينيكي باستهداف مسار الإشارات ذاتها على مستويات
متعددة. أما مزج العلاج الموجه بالعلاج المناعي، والكيميائي مع العلاجات الموجهة الأخرى فهو قيد البحث. وماذا عن العلاج المتعاقب، هل هو الأفضل؟
يمكن أن يكون (VEGF) يقول د. جيشوند إن الدلائل الحالية تفترض أن الاستخدام المتعاقب لمثبطات معامل نمو الجدار الداخلي للأوعية الدموية
وعليه فإن الفترة القصوى لتحقيق الفائدة الإكلينيكية .(mTOR) مفيدا من دون أن يعيق الاستجابة التالية لمثبطات مستهدف رابامیسین للثدييات
علاجات موجهة * أما عن دور * .(mTOR) قبل تحويل طريقه العمل إلى مثبطات (VEGF) يمكن أن تتم باستخدام اثنين من مثبطات
العلاجات الموجهة في علاج سرطان خلايا الكلى، فمع بداية القرن الـ21، تغير أسلوب علاج سرطان خلايا الكلى المنتشر بصورة جذرية نتيجة للمعرفة
يقاوم (mRCC) الثاقبة الجديدة لعلم البيولوجيا الجزيئية بالنسبة للأورام وأيضا لاختيارات العلاج الجديدة المتاحة. ولأن سرطان خلايا الكلى المنتشر
العلاج الكيميائي بصورة كبيرة، فقد تم استخدام السيتوكينات لعلاج الحالات المتقدمة. وعلى الرغم من ذلك فإن هذه العلاجات مؤثرة في عدد محدود من
المرضى فقط، وغير مناسبة لغالبية المرضى بسبب السمية. أول نوع من العلاجات الموجهة التي تم تصميمها لتثبيط مستقبلات التيروزين كيناز، التي يعتقد
أنها مهمة لنمو الورم والأوعية الدموية المغذية له، كان هو «سورافينيب»، الذي تم إجازته لعلاج سرطان خلايا الكلى المتقدم في 2005 بالولايات المتحدة
الأميركية، وفي الاتحاد الأوروبي عام 2006. وتم إجازة مثبط التيروزين كيناز «سنيتتينيب» في 2006، ومثبط مستهدف رابامیسین الثدييات
«تيمسيروليمس»، مثبط معامل نمو جدار الأوعية «بيفاسيزوماب» إلى جانب الإنترفيرون لعلاج سرطان خلايا الكلى المتقدم في 2007. وقد سألنا
البروفسور زيا كيركالي، من كلية طب جامعة دوكاز أيلول بإزمير في تركيا، كيف يمكن الاختيار مع وجود هذه الاختيارات الكثيرة؟ فأجاب: إن كل مريض
يختلف عن الآخر، وعلى ذلك فإن علاجا واحدا لن يفيد جميع المرضى. وأعطى مثالا على ذلك المرضى المسنين. وبناء على تلك الحقيقة، فإن الجمعية الدولية
لسرطان المسنين توصي بالآتي: «عند الأخذ في الاعتبار أنسب الأدوية للاستخدام مع مريض بعينه يجب الأخذ في الاعتبار بيانات السمية الخاصة لكل
علاج موحد على حدة، وأيضا وجود مرض مصاحب محدد». وعلى ذلك، فكل البيانات — من الدراسات غير العشوائية، والتحليلات الفرعية، وبرامج
الوصول للمرضى الموسعة، والدراسات بأثر رجعي، ودراسات الحالة، والخبرة الإكلينيكية، يمكن استخدامها لبناء الأساسات التي تدعم اتخاذ القرار الإكلينيكي.
* تحسين العلاج الفردي * يقول البروفسور جواكيم بيلمونت، من مستشفى الجامعة المستقلة ومستشفي ديل مار، برشلونة، إسبانيا: إن العلاج الفردي
الأمثل لكل فرد داخل مسار العلاج يمكن أن يتحقق عن طريق الأخذ في الاعتبار احتياج كل مريض على حده. ونهج التركيز على المريض، الذي تم وضعه
بواسطة فريق من الخبراء يضم أطباء المسالك البولية والأورام من ذوي الخبرة من جميع أنحاء أوروبا، صمم لكي يعين العوامل المحددة التي ينبغي أخذها في
الاعتبار عند اختيار العلاج الأمثل للمرضى كل على حدة. النظام الجديد يضع في الاعتبار كلا من مقياس مركز كيترينج التذكاري للسرطان
وتحليل أنسجة الورم، وعدد ومواضع انتشار الورم، عمر المريض، وحالة أداء المريض، والأمراض المصاحبة، والعوامل المرتبطة بالعلاج. .(MSKCC)

وباستخدام هذا الأسلوب، قامت لجنة من الخبراء باستعراض أحدث البيانات المتاحة لتحديد المجموعات الفرعية من المرضى الذين قد يستفيدون من العلاج بـ«سورافينيب»، الذين لا يتم عادة تمثيلهم في تجارب المرحلة الثالثة. وتم تحليل كل من نتائج المرحلة الثانية والثالثة من التجارب السريرية، وبرامج الوصول للمرضى الموسعة، والتحليلات الفرعية، ودراسات المراكز الفردية. هناك بيانات قوية تدعم استخدام «سورافينيب» لطائفة من المجموعات الفرعية للمرضى، بمن فيهم المسنون والمرضى الذين يعانون من الفشل الكلوي وتليف الكبد، ولديهم انتشار للورم في ما يصل إلى 3 أعضاء مختلفة من الجسم، ولديهم مقياس EU - ARCCS جيد أو متوسط وارتفاع ضغط الدم المعالج. * فترات حياة أطول * في تحليل مجموعة فرعية من بيانات الدراسة MSKCC كان متوسط فترة الحياة من دون تطور للمرض متماثل للمرضى فوق وتحت سن 70 عاما (23.9 مقابل 26.3 أسبوعا على التوالي). كما كانت تدعم استخدام «سورافينيب» مع المرضى الذين لا يناسبهم أو لا يستجيبون للعلاج بـ«السيتوكين، علاوة على قلة عدد الأحداث السلبية من الدرجات 3 - 4 التي تصاحب هذا العلاج في حالة كبار السن مقارنة بالأدوية الموجهة الأخرى. كما أظهر تحليل فرعي آخر للدراسة نفسها أن استفادة المسنين من العلاج هل العلاج (SWITCH) «كانت متقاربة مع الأصغر سنا، مع ميل لوجود فترة بقاء أطول من دون تطور. * دراسات قائمة * التجربة «سويتش كخط أول بـ«سورافينيب» أكثر فائدة من «سنيتينيب» في العلاج المتعاقب؟ سيتم الإجابة على هذا السؤال عن طريق المرحلة الثالثة العشوائية والمفتوحة هذه الدراسة بدأت في يناير (كانون الثاني) 2009، ويخطط لها الاستمرار حتى عام 2012. وتشمل 540 .(SWITCH) للتجربة الدولية مريضا. أهم خصائص إدماجهم في التجربة هي: أنهم مرضى سرطان خلايا الكلى المصحوب بانتشار للورم ولا يصلحون للعلاج بالسيتوكينات، الذين يكون الدواء بالنسبة إليهم هو الخط الأول للعلاج. الذراع الأولى، يتم إعطاؤهم «سورافينيب»، وبعد وقف العلاج بسبب تطور المرض أو ظهور أعراض جانبية يتم إعطاؤهم «سنيتينيب»، والعكس صحيح بالنسبة للذراع الثانية. الهدف الرئيسي هو تقييم ما إذا كانت فترة البقاء خالية من تقدم المرض من وقت الاختيار العشوائي حتى تطور المرض أو الوفاة. وإحدى النقاط التي تركز عليها الدراسة هي تحليل «سمية» عضلة القلب، التي تتم عن طريق رسم القلب .وتحليل مؤشر هبوط القلب الاحتقاني. د. عبد الحفيظ يحيى خوجه المصدر: الشرق الأوسط

4.6 Inventory of Words Containing Schwa

The number of words containing schwa appears very high in the corpus; this is because schwa is the most frequent vowel in English. Thus it occupies many positions; it is located in the first, mid and last position. The number of schwa in all positions is 28 as shown in Table 4.3. However, the number of schwa occurring in the last position is 24.

4.7 Inventory of Words Containing ẖ Sound in Final Position

It should be pointed out that the h sound of pause never occurs in the first position, so the selection of words emphasises only on words that contain /t/ in final position and replaced by /h/. The number of words that encompass h sound in the last position is 30. Table 4.4 shows clearly the selected words and their number, including phonetic transcription.

4.8 Inventory of Schwa and h Sound in Final Position

The number of English words containing final schwa is 24, while the number of Arabic words containing final h is 30.

Table 4.3 Words containing schwa in Arabic

Num	Word	PhoneticTrans
1	Potential	/pətenʃl/
2	Hospital	/hɒspɪtl/
3	Access	/əkses/
4	Problem	/prɒbləm/
5	Surgeons	/sɜ:ʤənz/
6	Total	/təutl/
7	Clear	/klɪə/
8	Disappear	/dɪsəpɪə/
9	Future	/fjuʧə/
10	Never	/nevə/
11	Computer	/kəmpjutə/
12	Honour	/Λnə/
13	Cure	/kjuə/
14	mucker	/m^kə/
15	Sacra	/ s ækrə /
16	Hour	/auə/
17	Hear	/hɪə/
18	Quarter	/qwɒtə/
19	Order	/ɔ:də/
20	Wire	/waɪə/
21	Under	/Λndə/
22	There	/ðeə/
23	Better	/betə/
24	Receiver	/resi:və/
25	either	/aɪðə/
26	together	/təgeðə/
27	House sitter	/sɪtɔ /
28	judder	/ʤ^ də/

4.9 Words in Contrast

As shown in Table 4.5, there are eight contrasted words. They are similar in terms of pronunciation but different in the graphical system and phonological system as well. In what follows, a contrastive study is made to show differences (Table 4.5).

Table 4.4 Pause located in the Arabic words

Number	Word	PhoneticTrans
1	فعالة	/faʕa:lah/
2	القاتلة	/alqa:tilah/
3	الخاصة	/alxa:sɪah/
4	الحاصلة	/alha:sɪlah/
5	العربية	/alʕarabɪah/
6	الإصابة	/alɪsɪa:bah/
7	المكرمة	/almukarama/
8	الشرقية	/aʃarqɪah/
9	العلمية	/alʕɪlmɪah/
10	الدوائية	/adawaʔɪjah/
11	الجمعية	/alʒamʕɪjah/
12	السعودية	/asaʕudɪjah/
13	الوقاية	/alwɪqa:jah/
14	الإشارة	/alʔɪʃa:rah/
15	سكرة	/Sakrah/
16	متكررة	/mutakarɪrah/
17	المرضية	/almaradɪjah/
18	للكلية	/lɪlkɪljah /
19	المنتشرة	/almuntaʃɪrah/
20	المستهدفة	/almustahdafah/
21	الموجهة	/almuwaʤahah/
22	المماثلة	/almuma:Θɪlah/
23	الخطورة	/alxutɪu:rah/
24	محددة	/muhadadah/
25	بالدراسة	/adɪra:sah/
26	التالية	/ata:lɪjah/
27	المائة	/almɪʔah/
28	الخلية	/alxalɪjah/
29	الواضحة	/alwa:dɪhah/
30	واحدة	/wa:hɪdah/

Table 4.5 Words in contrast

Sitter	**/sɪtə/**	**/sittah/**	ستة
Sacra	**/seɪkrə/**	**/sakrah/**	سكرة
Judder	**/ʤʌdə/**	**/ʒəddah/**	جدة
Mucker	**/mʌkə/**	**/makkah/**	مكة

4.9.1 Segmentation to Show Contrast

It is so important to mention that the use of segmentation is made for two purposes: to show equivalence of schwa in Arabic and to delimit the phonetic value. In doing so, the four aforementioned principles, i.e. linearity, biuniqueness, invariance and local determinacy, are used in both Arabic and English words.

4.9.1.1 Segmentation of the Word Sitter and ستة

Linearity: The word **sitter /sɪtə/** contains an alveolar voiceless fricative, a short vowel, an alveolar voiceless plosive/t/ and a schwa. As for the word **/sɪttaḥ/**, it contains an alveolar voicclcss fricativc, a short vowel, a repeated alveolar voiceless plosive /t/, a short front open vowel and a voiceless laryngeal fricative / **ḥ**/

Biuniqueness: Each phoneme corresponds to a phone.

sitter = /sɪtə /s=s,i=ɪ ,t= t , er = ə

ستة **= / sɪttaḥ/**س=/s/=//ɪ=ɪ/t/ = ت a=/a/ ه=/ḥ/

Inviariance: The word **sitter / sɪtə/** contains two syllables/ sɪ/and / tə/. The first syllable involves a voiceless alveolar consonant /s/ and a high mid front vowel, while the second one embraces an alveolar plosive consonant and a schwa vowel. The Arabic word **/sɪttaḥ/** encompasses also two syllables /sɪ/ and/ taḥ/; the first syllable contains a voiceless alveolar consonant /s/ and a high front vowel, and the second syllable is composed of an alveolar plosive consonant, a front open vowel and a laryngeal voiceless fricative consonant. However, once investigating both words using contrast between segments, it is noticeable that consonants are constant, but vowels are variant.

Local Determinacy: It should be pointed out that there is no influence of sounds upon neighbouring sounds in both s, i.e. s sound of English and s sound of Arabic.

4.9.1.2 Segmentation of the Word sacra and sakraḥ

Linearity: The English word **sacra =/seɪkrə/** has a sequence of an alveolar voiceless consonant, a vowel, a voiceless plosive palatal consonant, an alveolar approximant consonant and a schwa vowel. Again, as with the Arabic word **/sakraḥ/**, it is necessary to mention that it has a phoneme sequence somehow similar to the previous word. It is, then, composed of an alveolar voiceless consonant, a vowel, a voiceless plosive palatal consonant, an alveolar approximant consonant, a vowel and a voiceless laryngeal fricative.

Biuniqueness: Each phoneme corresponds to a phone

sacra =/seɪkrə/ s=s,a=eɪ ,k=k,r=r,a= ə

/ḥ/ **sacra / sakra ḥ**/ s=s. a=a. k=k. r=r. ه=/ḥ/

Invariance: Both of the Arabic and the English words are formed with two syllables and same consonant sequence. The difference lies in vowels, that is, one can detect the differences operating at the level of vowels and not consonants. Consequently, consonants are constant but vowels are variant.

Local Determinacy: The point about local determinacy focuses on the influence of sounds upon neighbouring sounds. As for the phoneme sequence of both words, it is observable that the approximant consonant /r/ is preceded by a voiceless plosive which alters the approximant voiced consonant into devoiced consonant.

4.9.1.3 Segmentation of the Word Judder and جدّة

Linearity: The word **/ʤʌdə/Judder** contains a voiced palatoveolar affricate consonant, a vowel, a voiced alveolar plosive consonant and a weak vowel (schwa). The word (جدة) **/ʒəddaḥ/**, however, comprises a voiced palatal fricative consonant, a vowel, a repeated voiced alveolar plosive consonant, a vowel and a voiceless laryngeal consonant.

Biuniqueness: Each sound has a symbol and vice versa.

جدة **= / ʒəddaḥ/**ج= /ʒ/,/v/ =**ə**,د =/d/, a=/a/ ة=/ḥ/

Judder /ʤʌdə/ J =ʤ, u=**ʌ** ,d=d, er =ə

Invariance: The English word Judder / **ʤʌdə/** contains two syllables **/ʤʌ/** and **/də/**. The first syllable consists of a voiced affricate consonant and a vowel, while the second syllable comprises a voiced alveolar plosive and a schwa. The Arabic word **/ʒəddaḥ/**, however, is made of two minimal syllables; the first is composed of a voiced palatal fricative consonant and a vowel, whereas the second comprises a repeated voiced alveolar plosive, a vowel and a laryngeal voiceless fricative. One investigating the segments of both words after using contrast, the most observable thing to note is that consonants are constant, while vowels are variable. / **ʒ/** and **/ʤ/**should not be taken as variable ; the selection is made on purpose.

Local Determinacy: Again there is no impact of the adjacent sounds on the vowels and no allophone is needed.

4.9.1.4 Segmentation of the Word Mucker and مكة

Linearity: The word **/mʌkə/ mucker** consists of a nasal bilabial consonant, a vowel, a voiceless velar plosive consonant and a weak vowel (schwa). The word (مكة) **/makkaḥ/**, however, contains a nasal bilabial consonant, a vowel, a repeated voiceless velar plosive consonants and a voiceless laryngeal consonant.

Biuniqueness: Each sound has a symbol and vice versa.

مكة **= / makkaḥ/**م= /m/=//a=/a/ , k=/k/ a=/a/ ة=/ḥ/

Mucker/ mʌkə/ m=m. u=**Λ** .k=k. er =ə

Invariance: The English word **Mucker /mʌkə/** is composed of two syllables **/mʌ/** and **/kə/**. The first syllable contains a bilabial nasal consonant and a vowel, and the second syllable is made of a voiceless alveolar plosive and a schwa. The Arabic word **/makkaḥ/**, in turn, has two minimal syllables; the first is made of a nasal bilabial consonant and a vowel, whereas the second comprises a repeated voiceless alveolar plosive, a vowel and a laryngeal voiceless fricative. The most important thing to note after inspecting the segments of the two words is that consonants are invariable, while vowels are fluctuant.

Local Determinacy: The neighbouring sounds have no effect on the vowels; as a result no allophone takes place.

4.9.2 Spectrum

It is very important to point out that vowels contain five formants represented in F1, F2, F3, F4 and F5 (Connor 1973 p. 71). The vowel /a/ is transcribed as follows [æ] and its F1 frequency is 731 Hz and its F2 frequency is 768 Hz. As for the F1 frequency of the vowel [i], it is estimated as 306 Hz, while the F2 is 2241 Hz. The F1 frequency of the [ɔ] however is 602 Hz and the F2 frequency is 884 Hz. Only the first and second formants are shown because they are considered necessary to delimit the quality of the sound according to acoustic phonetics (Sean 2011 p. 70). A spectrum is used in acoustic phonetics to decompose sound into frequencies; therefore each sound contains a different amount of energy in the different frequencies. It should be noted that this process can be compared to the white light decomposed into rainbow colours (Fry 1979 p. 182).

4.9.2.1 Spectral Analysis

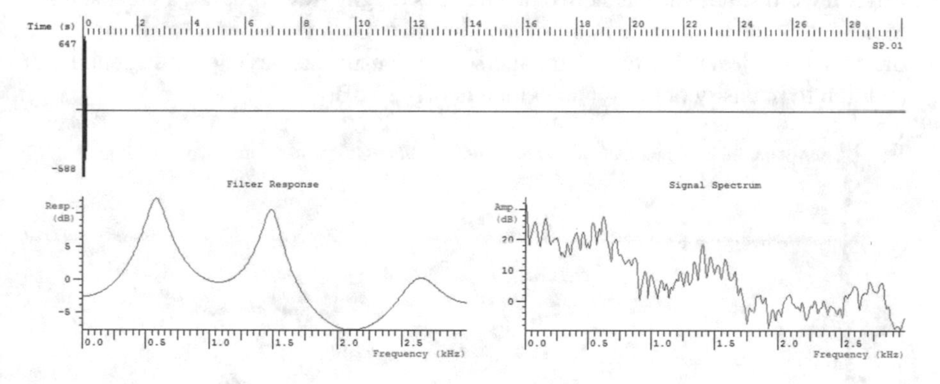

Tersitterang

F1 580 Hz **F2 1490 Hz** **F3 2638 Hz**

Amp **32 dB** Amp **29 dB** Amp **12 dB**

Int: 0.1620 dB

Per=0.000166

This diagram displays the signal of the last syllable located in the English word sitter. The spectrum on the right represents the signal of the spectrum where the peaks are not clear because it is not filtered. The spectrum in the left, however, contains very clear peaks and amplitude as well. It comprises three formants and three amplitudes. Formant 1 is located in the area of 580 Hz and 32 dB, formant 2 is found in the area of 1490 Hz and 29 dB and formant 3 is in the area of 2638 Hz and 12 dB.

Tersitterang
Wide Spectrogram

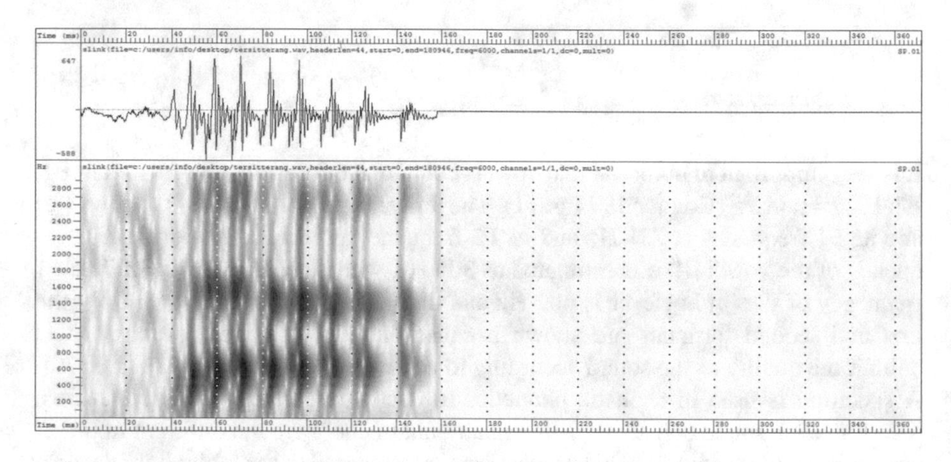

Syllable found in the English word "sitter"
F1 580 Hz F2 1490 Hz F3 2638 Hz
Intensity: 0.1620 dB

Reading off this spectrum that concerns the unit of the sound (ter) found in the English word sitter, one can notice that there is a light grey colour of a great surface representing the time space in milliseconds and a black colour of a small surface in the horizontal level that represents the sound density displaying fundamentals, in addition to intensity of the sound, which is 0.1620 dB.

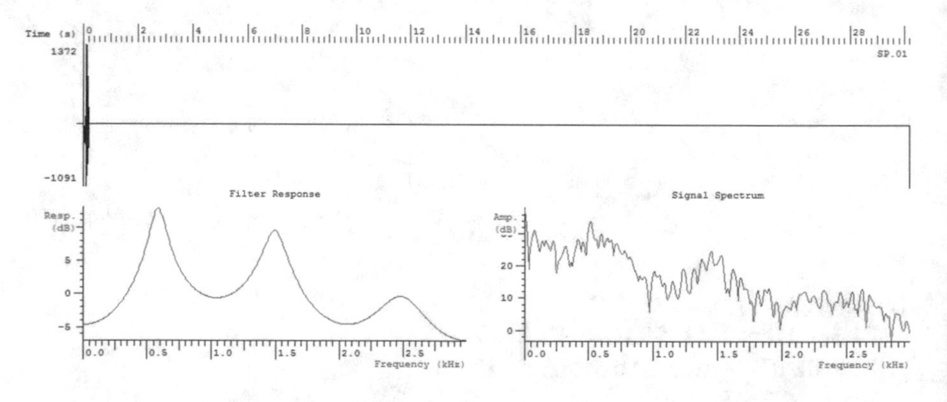

Tahsittahara
F1 583 Hz F2 1495 Hz F3 2489 Hz
Amp 32 Amp 29 Amp 12
Int: 0.3117 dB

Per=0.000166

The two diagrams represent the unit of sound of the Arabic word ستة. They are different from each other in terms of filtering, that is, the one located on the left appears very clear and comprises three formants, where the F1 is located in the area between 583 Hz and 32 dB, F2 in the area between 1495 Hz and 29 dB and F3 in the area between 2489 Hz and 12 dB.

Tahsittahara

Wide spectrogram

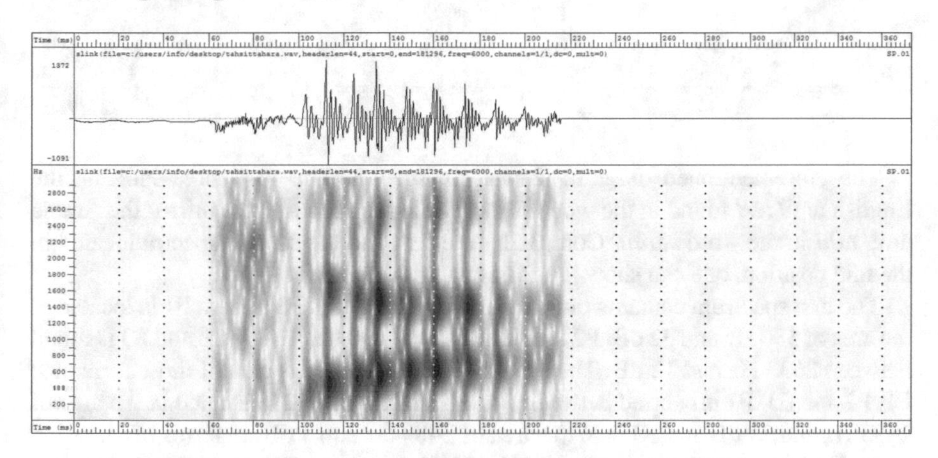

F1 583 Hz F2 1495 Hz F3 2489 Hz

Int: 0.3117 dB

Intensity of the Arabic phonological unit 'ستة' is 0.3117 dB as shown by the black colour represented by the sound density spread out in the grey surface on the horizontal level.

Tersitterang

F1 580Hz F2 1490Hz F3 2638Hz

Amp **32dB** Amp **29dB** Amp **12 dB**

Tahsittahara

F1 583 Hz F2 1495 Hz F3 2489 Hz

Amp 32 Amp 29 Amp 12

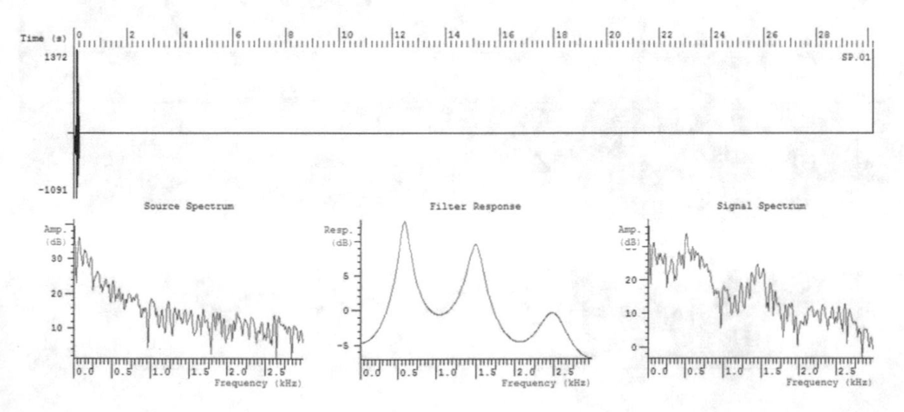

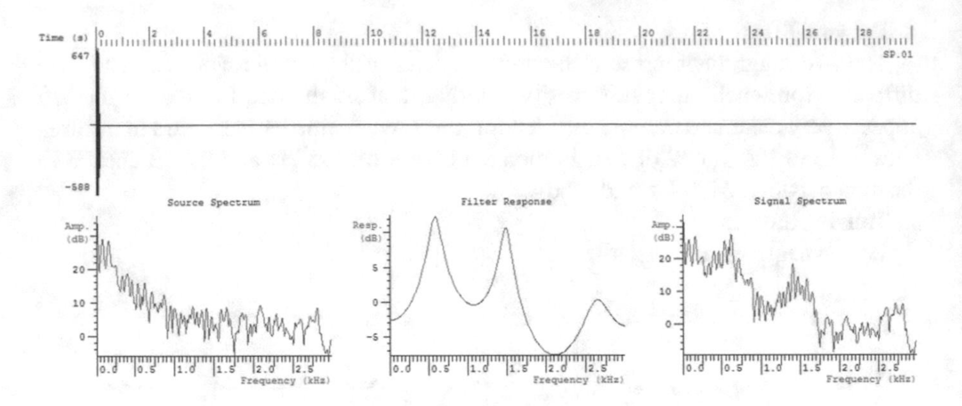

The aforementioned diagram shows six spectra. The first three indicate the English unit/ ***ter***/ found in the word **sitter**, while the second three mirror the Arabic unit/ ***tah***/ in the word ***sittah.*** Comparing the first and the second spectra located in the mid position, one can show the following:

The first spectrum contains three formants; F1, F2 and F3, where F1 is located in the area of 580 Hz and 32 dB, F2 is in the area of 1490 Hz and 29 dB and F3 is found between 2638 Hz and 12 dB. The second spectrum consists also of three formants: F1, F2 and F3. F1 is centred between 583 Hz and 24 dB, F2 is located in the area of 1495 Hz and 21 dB and F3 is in the area of 2489 Hz and 1 dB. In terms of intensity, a significant difference may be noticed with regard to the two units; English unit is 0.1620 dB, and by contrast the Arabic unit is 0.3117. The most overriding thing to mention is that both units have three formants with a slight difference in the frequencies; however the intensity appears to be different. The reason is that each language has its system regarding stress in other words; if an English word contains two syllables and the last syllable encompasses schwa, it is therefore not stressed. As a conclusion, the units are similar because the coefficient of correlation is 0.98.

Tersitterang
Wide spectrogram
Int :0.1620 dB

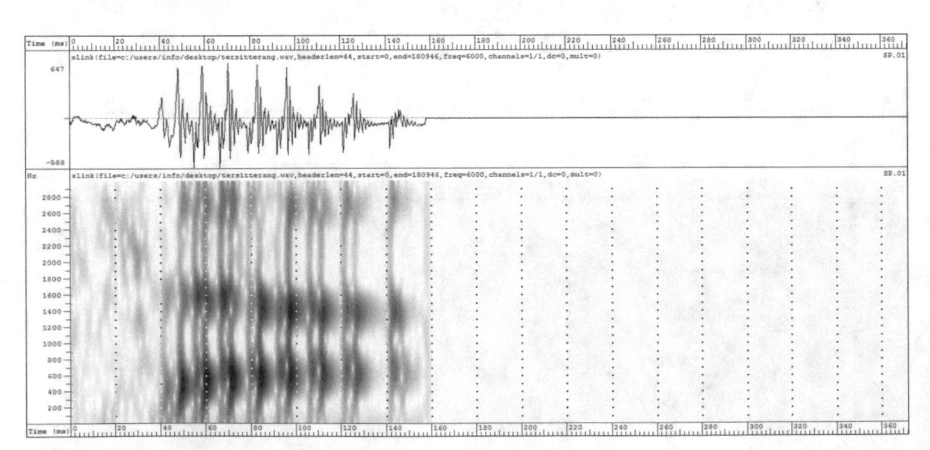

Tahsittahara
Wide spectrogram
Int : 0.3117 dB

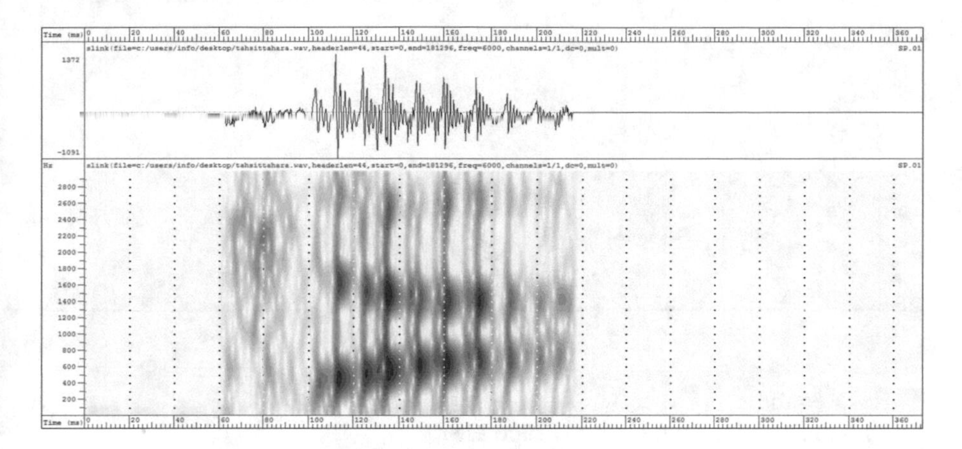

(Tersitterang(Int: 0.1620 dB
(Tahsittahara)Int: 0.3117 dB

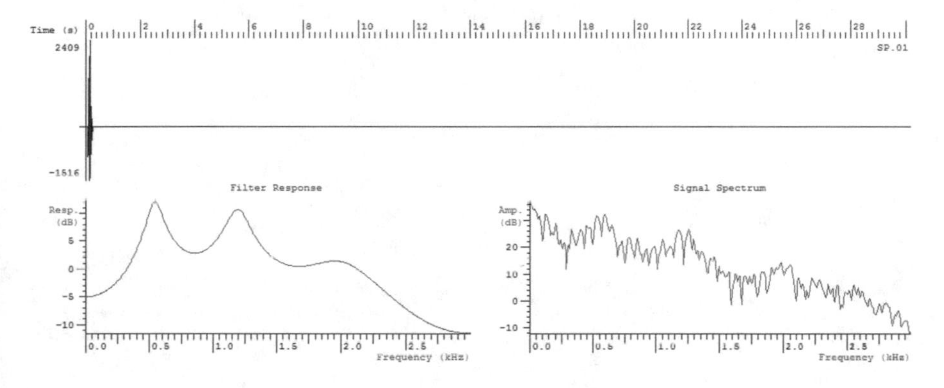

Rasacraang
F1 540 Hz F2 1196 Hz F3 2006 Hz
Amp 32 dB Amp 29 dB Amp 12dB
Int: 0.1283 dB
Per=1/6000=0.0001667
0.2393s time duration

The phonological unit **/ra/** of the English word (**sacra**) as displayed by the aforementioned diagrams contains three formants: formant 1 is located in the area of 540 Hz and 32 dB, formant 2 is in the area of 1196 Hz and 29 dB and formant 3 is in the area of 2006 Hz and 12 dB.

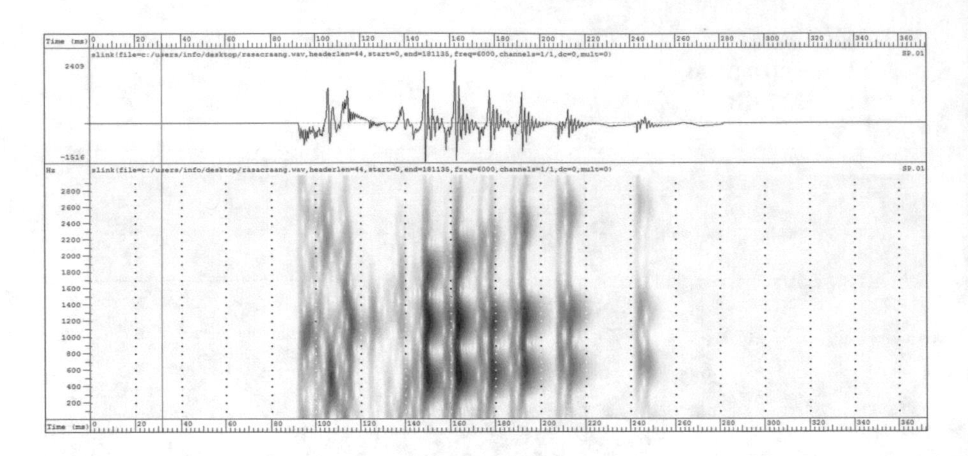

Rasacraang
F1 540 Hz F2 1196 Hz F3 2006 Hz
Int: 0.1283 dB

Intensity of the Arabic phonological unit is 0.1283 dB. It can be noticed from the sound density, which appears in black colour spread out in the grey surface.

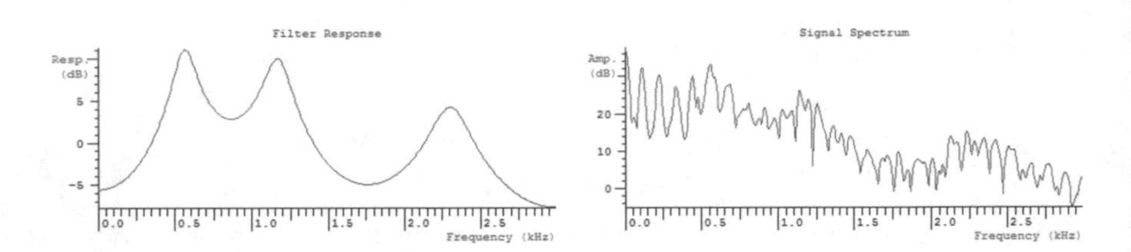

Rasacrahara
F1 541Hz F2 1084 Hz F3 2307 Hz
Amp 34 Amp 22 Amp 12
Int: 0.2884 dB
Per 0.000166

The unit ***/ra/*** in the Arabic word sacra contains three formants as shown in the spectrum mentioned above. The first formant is located in the area of 541 Hz and 34 dB, the second formant is in the area of 1084 Hz and 22 dB and the third formant is in the area of 2307 Hz and 12 dB.

Rasasacrahara
Wide spectrogram

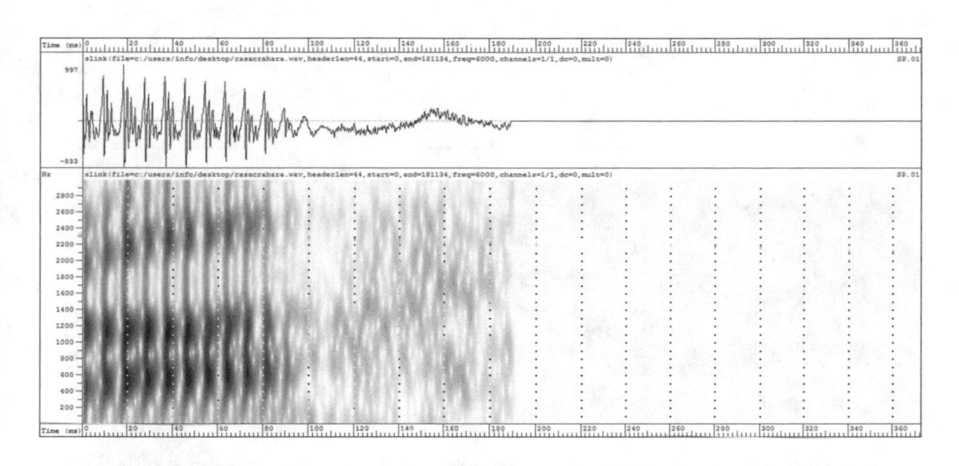

Rasacrahara

F1 541Hz F2 1084 Hz F3 2307 Hz

Int: 0.2884 dB

The intensity of the Arabic unit **/ra/** in the word **sacra** is 0.2884 dB; this can be shown from the sound density in the black colour spread out over the grey surface in the horizontal level.

Rasacraang

F1 **540 Hz** F2 **1196 Hz** F3 **2006 Hz**

Amp 32 Amp 29 Amp 12

Rasacrahara

F1 **541Hz** F2 **1084 Hz** F3 **2307 Hz**

Amp 34 Amp 22 Amp 12

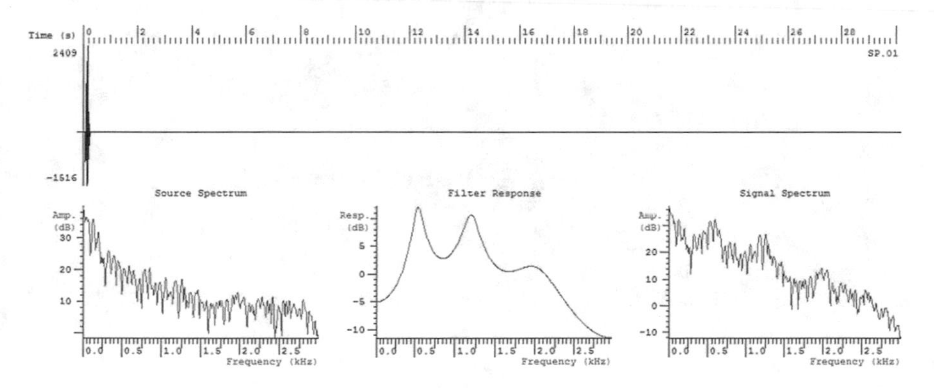

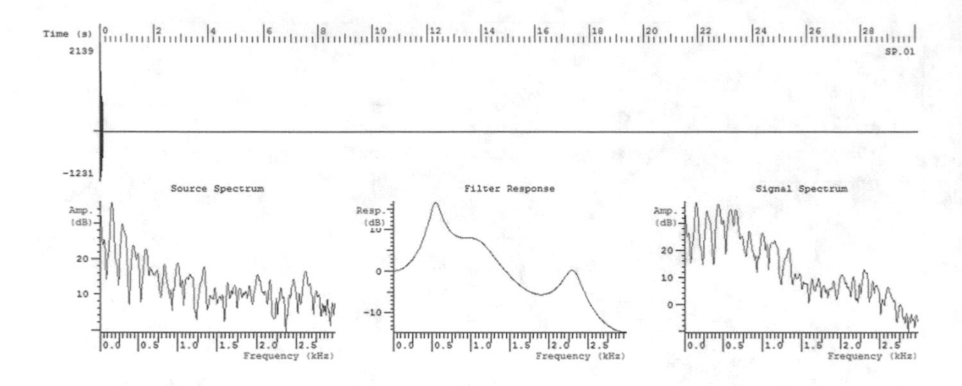

To compare the two phonological units of Arabic and English hinging upon the spectra mentioned above, we can say the following: the Arabic unit ***/rah/*** in the word ***sacrah*** and the English unit ***/ra/*** in the word **sacra** have three formants: F1 of both units are located in the area of 540 Hz and 32 dB and 541 Hz and 34 dB; in contrast, F2 and F3 of both units are different, in addition to intensity which is estimated to be 0.1283 dB in the English unit and 0.2884 dB in the Arabic unit, as shown in the aforementioned spectra. This is due to the degree of stress which is strong in Arabic and weak in English, especially in syllables that contain schwa. In terms of coefficient of correlation, the rate is 1.53, which indicates that the units are similar in the first formants.

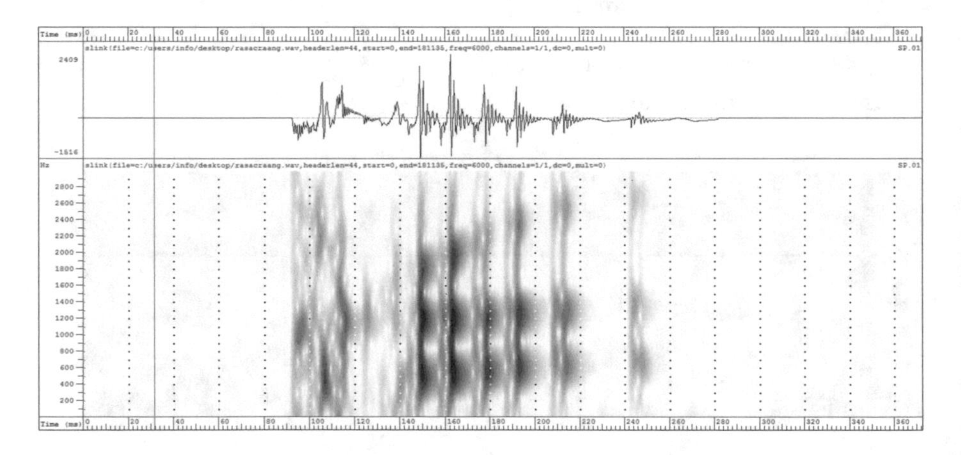

Rasacraang
F1 540 HZ F2 1196 Hz F3 2006 Hz
Int : 0.2884 dB

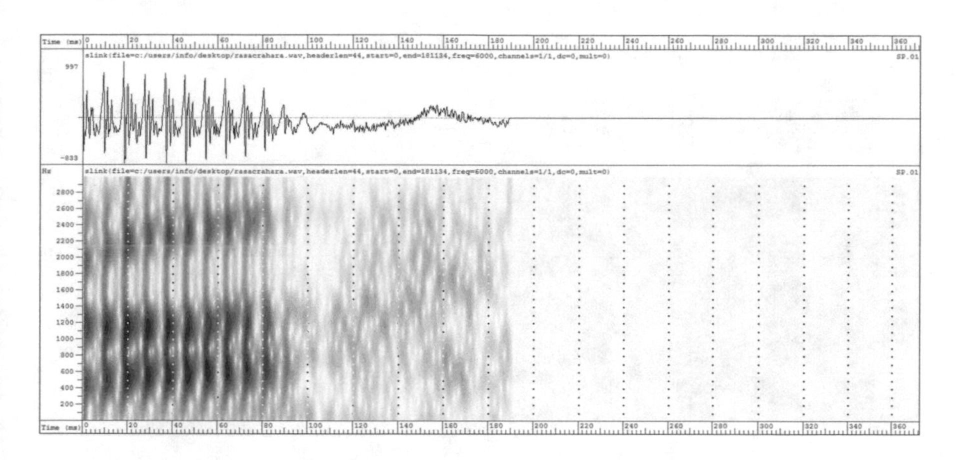

Rasacrahara
F1 541Hz F2 1084 Hz F3 2307 Hz
Int : 0.1283 dB
derjudderang

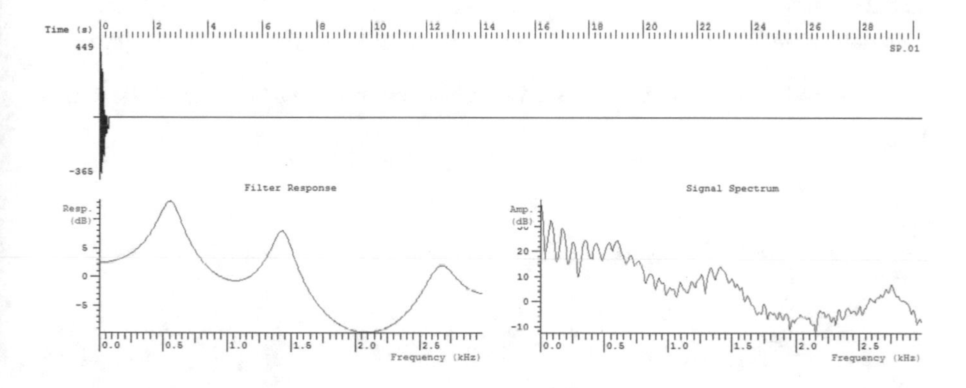

F1 558 Hz F2 1436 Hz F3 2667 Hz
Amp 22 Amp 13 Amp 9
Int: 0.2795 dB
Per=1/6000=0.0001667

These spectra represent the English phonological unit (***der***) in judder. And as all the other aforementioned spectra, we notice two types: one filtered and the other not. It is also noticeable that the clear spectrum shows three formants, where F1 is found in the area of 558 Hz and 22 dB, F2 is between 1436 Hz and 13 dB and F3 is between 2667 Hz and 09 dB.

derjudderang
Wide spectrogram

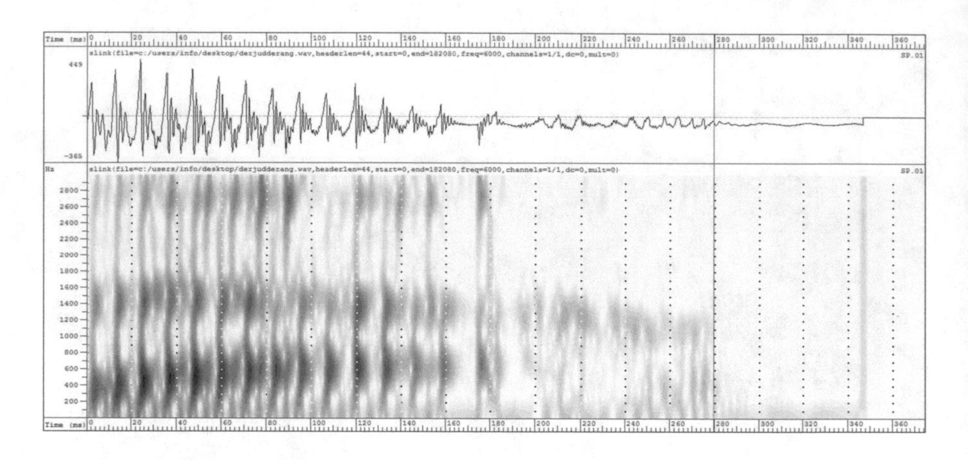

F1 558 Hz F2 1436 Hz F3 2667 Hz
Int: 0.2536 dB

This diagram displays the intensity of the phonological unit ***/der/*** in **judder**, which appears in the spread out of the black colour in the horizontal level and is **0.2536 dB.**

dajaddahara

F1 541 Hz F2 1359 Hz F3 2576 Hz
Amp 14.5 Amp 12 Amp 2.5
Int : 0.2795 dB

The Arabic phonological unit ***/dah/*** in the word ***jaddah*** represented in the spectrum posited in the middle displays three formants: F1 is located between 541 Hz and 14 dB, F2 is between 1359 Hz and 12 dB and F3 is in the area between 2576 Hz and 2.5 dB.

Dajaddahara
Wide spectrogram

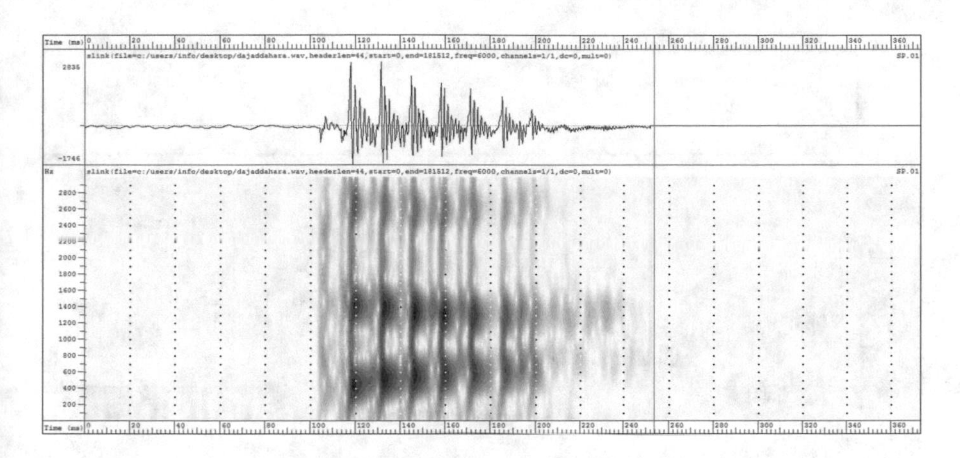

F1 558 Hz F2 1436 Hz F3 2667 Hz
Int :0.2795 dB

The aforementioned diagram displays three black lines that represent formants, in addition to the spread out of black colour over the grey surface, which indicates the density of the sound estimated to be 0.2795.

Derjudderang

F1 **558 Hz** F2 **1436 Hz** F3 **2667 Hz**
Amp 22 Amp 13 Amp 9

Dahjaddahara

F1 **541 Hz** F2 **1359 Hz** F3 **2576 Hz**
Amp 14.5 Amp 12 Amp 2.5

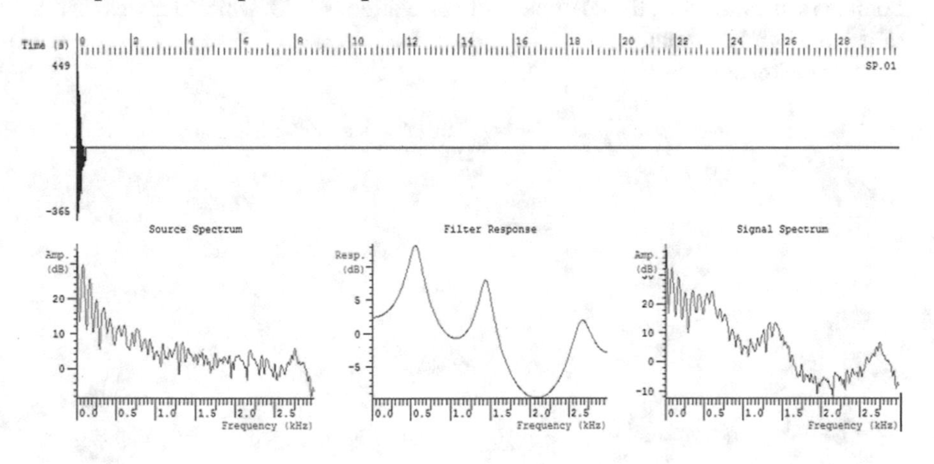

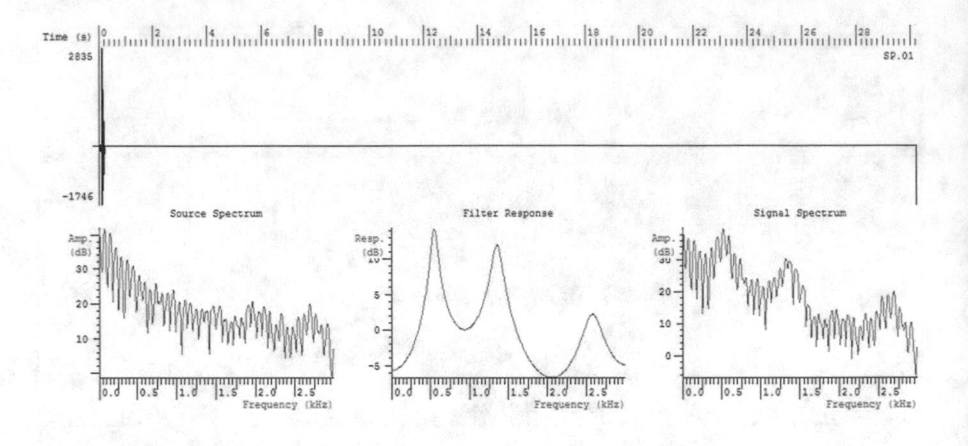

Reading off the above spectra that concern the two phonological units, i.e. Arabic unit ***/dah/*** in **Jeddah** and English unit ***/der/*** in **judder**, one can notice that both spectra contain three formants. Formant 1 of the English unit is centred in the area of 558 Hz and 22 dB, formant 2 is in the area of 1436 Hz and 13 dB, while formant 3 is in the area of 2667 Hz and 9 dB. As for the Arabic unit, it comprises also three formants with a slight difference in terms of values of frequencies compared to the English unit. So, F1 can be seen in the area between 541 Hz and 14.5 dB, F2 in the area of 1359 Hz and 12 dB, while F3 in the area of 2576 Hz and 2.5 dB. In terms of intensity, the remarkable thing is that the Arabic unit has a greater value (0.2795 dB) than the English one (0.2536 dB), which is due to the secondary stress which falls on the last syllable compared to English syllable, which is not stressed because it contains schwa. Finally, the coefficient of correlation is 1.53, which denotes that the units are slightly similar.

kermuckerang

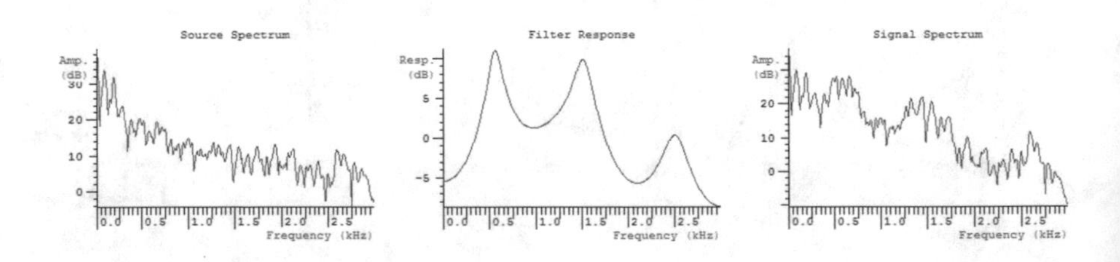

F1 **557 Hz** F2 **1173 Hz** F3 **2507 Hz**
Amp 12 **Amp 9.5** **Amp-1**
Int : 0.2220 dB
Per=1/6000=0.0001667

Again, if we try to interpret the spectrum of the English phonological unit (***ker***) in the word mukker, posited on the middle, which seems clearer than in the other two spectra, we may notice that it comprises three formants. The first formant is located in the sphere between 557 Hz and 12 dB, the second formant is between 1173 Hz and 9.5 dB, while the third formant is in the area of 2507 Hz and −1 dB.

Wide spectrogram

kermuckcrang

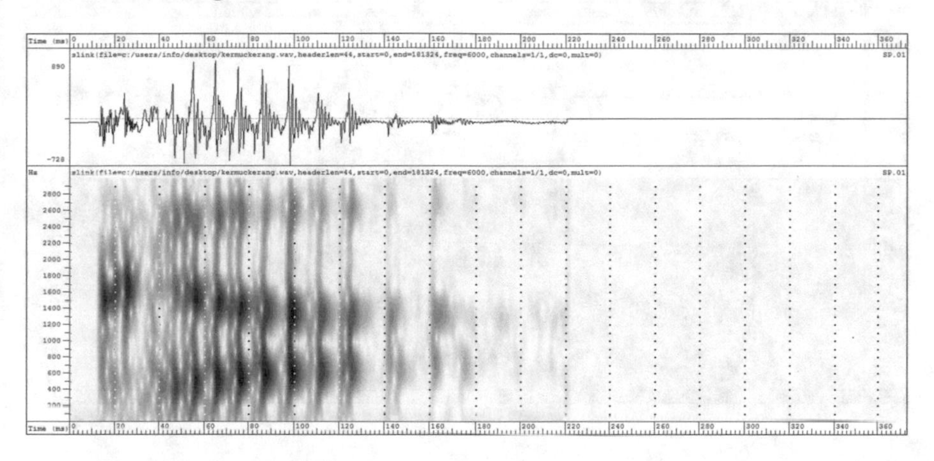

F1 557 Hz F2 1173 Hz F3 2507 Hz

Int : 0.2220 dB

This diagram shows intensity of the English phonological unit (***ker)*** in mukker in black, spread out over the grey surface and estimated at 0.2220 dB.

Kahmakkahara

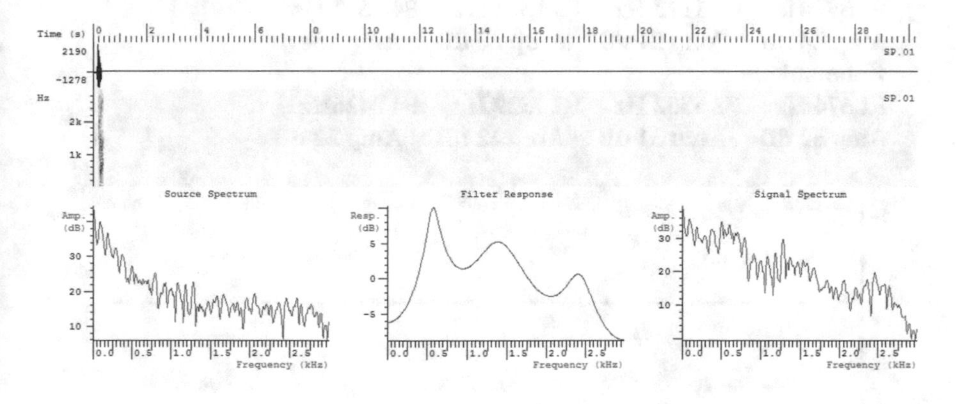

F1 574 Hz F2 1352 Hz F3 2426 Hz

Amp 10.5 dB Amp 5.5 Amp 1

Int : 0.3200 dB

The spectrum in the middle represents the Arabic phonological unit (***Kah***) in ***makkah***; it comprises three formants. The first formant is located in the area of 574 Hz and 10.5 dB, the second formant is between 1352 Hz and 5.5 dB and the third formant is between 2426 Hz and 1 dB.

kahmakkahara

Wide spectrogram

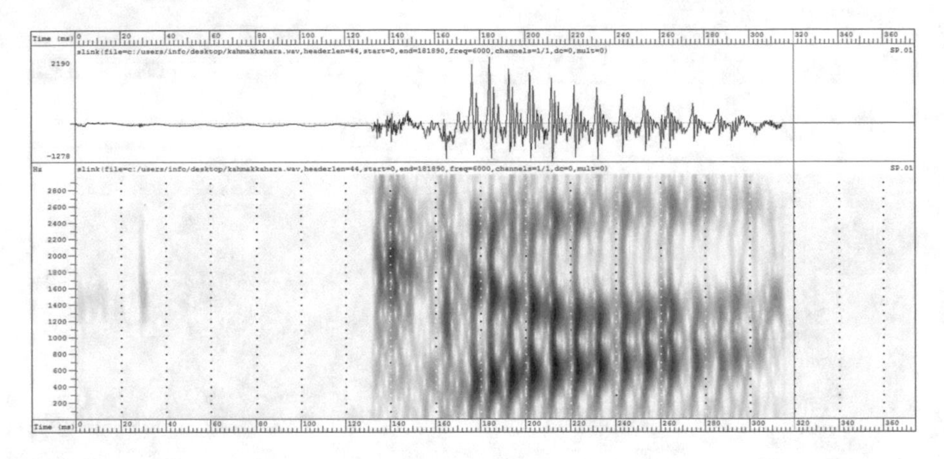

F1 574 Hz F2 1352 Hz F3 2426 Hz

Int : 0.3200 dB

As displayed by the diagram mentioned above, the intensity of the Arabic phonological unit is estimated at 0.3200 dB. This is from the sound density of black colour spread out over the grey surface.

Kermuckerang

F1 **557 Hz** F2 **1173 Hz** F3 **1521 Hz** F4 **2507 Hz**

Amp 24 dB Amp 21 dB Amp 20 dB Amp 13dB

Kahmakkahara

F1 **574 Hz** F2 **1352 Hz** F3 **1529 Hz** F4 **2426Hz**

Amp 32 dB Amp 31 dB Amp 22 dB Amp 22 dB

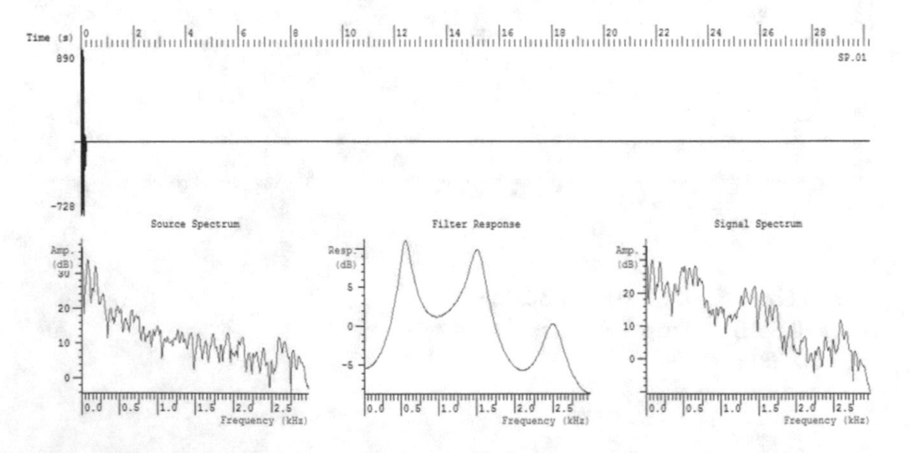

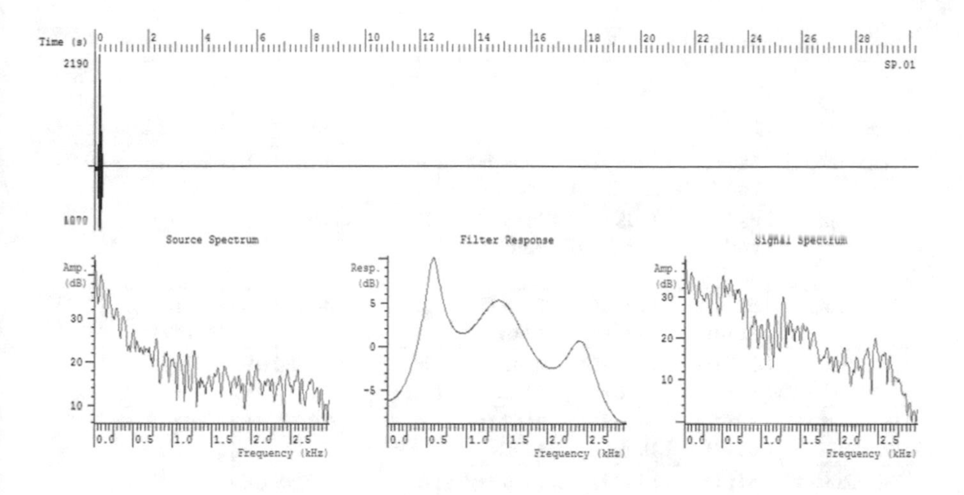

The comparison of the two phonological units in (***mukker***) and (**makkah**) may show the following: both spectra contain four formants. The first formant of the English unit is centred between the 557 Hz and 24 dB, the second is between 1173 Hz and 21 dB, the third is between 1521 Hz and 20 dB, while the fourth is between 2507 Hz and 13 dB. As for the spectrum of the Arabic unit, its first formant is located in the area between 574 Hz and 32 dB, its second formant is between 1352 Hz and 31 dB, its third formant is between 1529 Hz and 22 dB, while its fourth is located between 2426 Hz and 22 dB. In terms of intensity the Arabic unit is 0.3200 while the English unit is 0.2220. As a result we may say that both spectra contain four formants with a slight difference in the rate of frequencies and intensity which is due to the different systems of stress in Arabic and English. As for the coefficient of correlation, it is estimated at 0.7; this means that the two spectra are somehow similar in the formants and different in intensity. Table 4.6 can summarize all the details regarding the contrasted units; it provides numerical values about formants, the intensity, rate of similarity and average of coefficient.

4.10 Conclusion

In a summary, then, this chapter contrasts the vowel schwa with the /h/ sound of pause located in the last position of the words. To this effect phonemic taxonomy and a spectral analysis are applied. It should be noted that phonemic taxonomy applies four principles, linearity, biuniqueness, invariance, and local determinacy, and spectral analysis, where similar words are taken from two corpora (English and Arabic) and are segmented according to phonemic taxonomy, then recorded the last units of the similar words produced by different people in GoldWave (Speech Analyzer), then processed by the Sound Filing System (SFS), and finally exhibiting the results obtained.

Table 4.6 Spectra of contrasted units

Signal	F1	F2	F3	F4	Int	Rate of sim	Aver of coef
Tersitterang	**580 Hz and 1 dB**	**1490 Hz and 21 dB**	**2338 Hz and 1 dB**	**X**	**0.1620 dB**	**Complete sim**	**0.98**
Tahsittahaara	**583 Hz and 32 dB**	**1495 Hz and 29 dB**	**2489 Hz and 12 dB**	**X**	**0.3117 dB**		
Rasacraang	**550 Hz 32 dB**	**1196 Hz 29 dB**	**2006 Hz 12 dB**	**X**	**0.1283 dB**	**Sim in the first two formant**	**1.53**
Rasacraara	**541 Hz 34 dB**	**1084 Hz 22 dB**	2307 Hz 12 dB	**X**	**0.2884 dB**		
Derjudderang	**558 Hz 22 dB**	**1436 Hz 18 dB**	**2667 Hz 11 dB**	**X**	**0.2536 dB**	**Comp sim**	**0.7**
Dajaddahara	**541 Hz 35 dB**	**1359 Hz 30 dB**	**2576 Hz 19 dB**	**X**	**0.2795 dB**		
Kermukerang	**557 Hz 24 dB**	**1173Hz 21 dB**	1521 Hz 20 dB	**2507 Hz 13 dB**	**0.2220 dB**	**Comp sim**	**0.1**
Kahmakkahara	**574 Hz 32 dB**	**1352 Hz 31 dB**	**1529 Hz 22 dB**	**2426 Hz 22 dB**	**0.320 dB**		

References

Chomsky N (1970) Current issues in linguistic theory, 5th edn. Mouton, The Hague

Connor OJD (1973) Phonetics a simple and practical Introduction to the nature and use of sound in language. Cambridge University Press, Cambridge, Cambridge new edition. ISBN 014 013638x

Fry DB (1979) The physics of speech, Cambridge textbooks in linguistics, 1st edn. Cambridge University Press, Cambridge

Lass R (1984) Phonology an introduction to basic concept, 1st edn. Cambridge University Press, Cambridge

Sean AF (2011) Speech spectrum analysis signals and communication technology. Springer, Berlin. ISSN 1860-4862

Chapter 5
Arabic Automatic Speech Recognition

5.1 Introduction

Language is a complex phenomenon of multiforms, for it is associated in many fields; sometimes it is acoustic, sometimes physiological and sometimes psychological and social (De Saussure 2002, p. 15).

One of the most principal characteristics of language is the human ability to transmit complicated messages of daily activities with all members of society (Gimson 1980, p. 34) and to communicate during one's whole life without paying attention to how complex speech phenomenon is.

Speech production hinges first and foremost on exhalation; lungs eject carbon dioxide into the atmosphere; it goes through the larynx where the vocal folds dwell. They are separated by a space which is adjusted by laryngeal muscles. When the vocal cords are closed, receiving a pressure from the lungs, there will be voicing or phonation.

5.2 Speech Chain

In order to understand what goes on in the process of speech, it is necessary to look at two persons talking to each other; one is the speaker and the other is the listener. The speaker sends information, whereas the listener receives them. The chain of events used in sending information is called the speech chain, where the speaker first organizes ideas in his brain, takes the decision about the speech that he will produce and then puts ideas into a linguistic form, selecting necessary words and sentences according to the system of each language. This operation is associated to the work of the speaker's brain, which is represented in electrical commands coming from cells and directed to muscles of phonetic apparatus (tongue, lips, jaw and

M. Dib, *Automatic Speech Recognition of Arabic Phonemes with Neural Networks*, SpringerBriefs in Applied Sciences and Technology,
https://doi.org/10.1007/978-3-319-97710-2_5

vocal cords). The electrical signal coming from cells pushes production of slight variable pressures to the air that surrounds the mouth, which travel through the air as sound waves and reach the listener's ear and stimulate auditory mechanism which in turn produces electrical signals that pass through acoustic cells to arrive at the listener's brain (Goldberg and Riek 2000).

As soon as electrical signals arrive at the brain through acoustic cells, the neural activity found in the brain is increased, thanks to electrical signals of the ear; this change of brain's activity is helpful for the recognition and comprehension of the message sent by the speaker.

In a broad view, this analysis tries to show how the process of speech starts at the linguistic level with regard to the speech chain in the speaker's brain, selecting appropriate words and sentences and how it ends at the psycholinguistic level in the listener's brain, which decodes the neural activity made throughout the acoustic cells (Goldberg and Riek 2000).

5.2.1 Speech Technology

Speech technology appeared in 1945–1949 in Sweden, then moved to the United States at MIT (Massachusetts Institute of Technology) and then came back again to Sweden in 1951. It is the result of a set of different sciences enabling speech communication field and technical applications (Fant 2004). These sciences contain acoustic phonetics, electronics, linguistics, phonetics, psychology, physiology and the theory of information. All this has contributed to the invention of the computer.

Much interest was given during the last two decades to the process of child language comprehension and speech production; in this respect many studies shed light upon the process of language comprehension; in general, the best way to highlight this topic is the distinction between organized phases in child language comprehension according to the chronological development.

5.2.2 Discrimination Between Speech Sounds

Eimas et al. (1971, cited in Goldberg and Riek 2000) showed that in their first month of birth, children not only can hear speech sounds but also can distinguish between sounds; they can, for example, differentiate between /p/ and /b/.

Comprehension of Words. From the sixth month to the twelfth month, infants start to understand words.

Production of babble. During the sixth to the twelfth month, the selection of sounds begins, i.e. infants start to pronounce strings of sounds (alternation of consonants and vowels), such as babababa. This phase for a deaf infant starts 6 months later (Oller and Eilers 1988, cited in Goldberg and Riek 2000).

Production of words. From the twelfth month to the fourteenth, the infant starts to produce his own first words (Benedict 1979, cited in Goldberg and Riek 2000); thus taking interest in his own environment has also reached the point that thoughts and feelings are directed towards very small and exact things to people and materials and the infant is also able to discover that the produced sounds have specific meanings (Oller and Eilers 1988, cited in Goldberg and Riek 2000).

5.3 Speech Recognition

Speech organs are considered biological because their primary function is to make the body survive. Their secondary function is speech. However, speech is different from one speaker to another, because of the many factors which influence these organs. They can be listed as follows: gender (male or female), education (environment), psychological situation, pronunciation, syllabification, nasalization and rate of speaking, in addition to some other external factors, such as the use of mobile phones in transmitting messages to people in distant localities. Consequently, all these factors complicate speech recognition compared to speech generation (Tebelskis 1995, p. 4).

Speech recognition is a very thorny area because to achieve a good result requires some conditions for evaluation. Contrary to some general conditions where it is impossible for the human brain to attain exactness, under some particular conditions, the human brain may reach exactness. The following characteristics have an impact upon evaluation conditions:

vocabulary size, isolated speech, discontinuous speech versus continuous speech, language constraints, text reading versus spontaneous speech, and adverse conditions.

Speech recognition is synonymous with intelligibility, i.e. a group of overlapping sounds or a word uttered by a speaker can be recognized by the listener. And, despite the production of a sentence or a word from the same speaker, acoustic segments are different, which may cause a problem. This is due to the different physical structures of speakers (men, women, children), which cause differences in the acoustic segments, because speech recognition is the rendering of a variant stimulus to a constant response, or the transformation of an analog signal into an acoustic signal, or the conversion of a digital signal into classified decisions (Plomp 2002, p. 94 cited in AWEJ V5N4, December 2014).

Accordingly, speech recognition can be viewed as trouble of communication between man and machine; in other words, the machine tries to identify a string of a word uttered by a speaker, whose speech generation process is intricate and goes through some stages; the brain first produces the text, which is composed of a sequence of sounds, then transformed into acoustic signals and then converted into audible waveform (Wu Chou 2009, p. 14 cited in AWEJ V5N4, December 2014). In other words, an analog signal is transformed into an acoustic signal or algorithmic one to obtain sorted outcomes or a changeable input is modified into a constant output (Plomp 2002 cited in AWEJ V5N4, December 2014).

Speech recognition is an inverse operation which starts from the speech waveform and ends in decoding a message. It is a mechanism able to decipher a speech signal coming from the vocal tract or nasal cavity and is represented in a sequence of linguistic units found in the message that the speaker wants to transmit (Rabiner and Juang 1993, p. 4).

The ultimate target of speech recognition is the ability to communicate with the machine. This interaction has witnessed many applications because of the fast progress of devices and technological programmes (Peinado and Segura 2006, p. 1 cited in AWEJ V5N4, December 2014).

Automatic speech recognition necessitates proficiency in the following areas: signal processing, acoustic phonetics, patterns recognition, communication, theory of information and physiology. In what follows a clear explanation of the human brain work (physiology) and computer tasks (programming) is given.

It is very interesting to know that a human being can recognize speech very easily; what is the reason behind this? The answer to this question requires comparison between human brain and computer. The structure of the human brain differs from the structure of the computer significantly, that is, the human brain works with some patterns different from those of the computer. Computers use very complex and rapid central processors with the instruction of a clear programme and an addressable memory, while the human brain uses an associative memory and extended number of cells called neurons that are densely connected with each other with synapses whose power changes by experience and accepts many constraints (Tebelskis. 1995, p. 9).

5.3.1 Fundamentals of Speech Recognition

Automatic recognition is a multilevel pattern recognition process. Acoustic signals are built into a hierarchy of units, such as phonemes, word phrases and sentences, and therefore each level provides extra temporal constraints. This hierarchy of constraints can be well exploited in combining decisions using a probabilistic operation in all levels from the bottom level to the top one (Rabiner and Juang 1993, p. 4).

5.4 Intelligence

The word intelligence is a vague term; however some definitions tried to elucidate the point. It is the aptitude to understand and to think (*Oxford Advanced Learner's Encyclopedia* dictionary). It is the ability to learn, comprehend and think (Longman Advanced Dictionary). There exist numerous sorts of intelligence, which can briefly be worded into seven types: linguistic intelligence, mathematical and logical intelligence, spatial intelligence, musical intelligence, kinaesthetic bodily intelligence, personal intelligence and interpersonal intelligence (Douglas 2007, p. 10 cited in AWEJ V5N4, December, Arab World English Journal 2014).

5.4.1 Artificial Intelligence

Artificial intelligence is the replication of all sorts of human intelligence through artificial means, such as neural networks. It aims at simulating human intelligence. The field of artificial intelligence grew out of tremendous areas such as the medical field and observing the tasks done of the human body and the brain in particular, linguistics and understanding the mechanism of language and as a result treat natural languages automatically; Natural Language Processing (NLP) (Dinedane 1995, p. 13 cited in AWEJ V5N4, December 2014). Moreover, artificial intelligence tries to realize tasks similar to that used by human intelligence. It entails robot behaviour, language comprehension, pattern recognition and knowledge representation (The Hutchinson Encyclopedia cited in AWEJ V5N4, December 2014).

5.4.2 Neural Networks

To understand the human brain's work is a challenge. There is no doubt that the best way which can allow us to examine data processing in the human brain mathematically and computationally is the use of the NNW for imitation or reproduction of tasks of human brain. NWW can be defined as a system which hinges upon mathematics, whose structure is made up of processors similar to the brain's cells. They encompass a set of nodes that collect input from various sources and then transmit them to other nodes, which in turn send them to other nodes. They can receive a very convoluted input and represent them in a very simple output. They comprise three layers: input layer, output layer and a hidden layer in between; each processor interacts with one another throughout synapses (Douglas 2007 cited in AWEJ V5N4, December 2014). They typify many human brain's functions, such as comprehension, calculation and memorization.

Neural networks are of three sorts; yet they share identical basic features listed as a set of processing units, a set of communication, a computing procedure and a training procedure. They have the power to transform a very complex input into a very simple output (Tebelskis 1995 cited in AWEJ V5N4, December 2014).

NNW represents many mathematical models of the human brain's functions, such as comprehension, calculation and memorization. It is a challenge in modern life to understand the process of the human brain's work, and undoubtedly, the best way which can enable us to investigate data processing in the human brain mathematically and computationally is the modelization of the NNW (Tang et al. 2007a, b, p. 128).

5.4.3 Fundamental of Neural Networks

There are many different types of NNW, but they are similar in terms of the same four basic attributes:

- A set of processing units
- A set of communication
- A computing procedure
- A training procedure (Tebelskis 1995, p. 28).
- Neural networks contain many simple processing units similar to those found in the human brain. These units work at the same time on the basis of some characteristics. In what follows a full description of how neural networks operate is given.

(a) ***Trainability***: Networks can be trained to link any form of input and output, for example, to classify linguistic patterns into phoneme category.
(b) ***Nonlinearity***: One of the most important characteristic of networks is the ability to calculate the different directions of input and allow them to execute random intricate modification of data; therefore they are practical for speech since it is of various directions.
(c) ***Robustness***: Networks can treat turbulent data which can help tremendously the networks to form better generalizations. And since speech consists of noisy patterns, generalization may contribute to a good treatment.
(d) ***Uniformity:*** Networks have the characteristics of providing a constant computational model whose job is to combine limitations from various inputs. In another way, to utilize the first inputs of referential speech and inputs of differential speech, for instance, synthetize acoustic and visual inputs in a multimodal system.
(e) ***Parallelism***: Networks are parallel and are useful for practice on densely parallel computers. This will permit very fast processing of speech or other data (Tebelskis 1995).

Automatic recognition is to identify patterns of various grades. For instance, acoustic signals are inspected, then converted into a group of units such as phonemes, words, phrases and sentences and, finally, are presented into numerical values; thereby, this scale of constraint can work perfectly in combining the decisions using mathematical probabilities in all levels and come out with final decisions (Tebelskis 1995).

In order to simulate the human brain and its functions, man attempted to invent machines able to produce and recognize speech, i.e. smart to decode signals loaded with a string of linguistic units found in the message of a speaker.

5.4.4 Neural Networks as a Set of Processing Units

Neural networks contain several simple units of processing that resemble brain cells. These units work at the same time relying on parallelism and do mathematical operations (The Hutchinson Encyclopedia). Units are divided into input, which receives data from environment and transfers its representation into output in the form of decisions or signals (The Hutchinson Encyclopedia).

As regards the architecture of NNW, it should be noted that units are represented through circles and there is a consensus that input units are located in the upper layer, while the output units are found in the lower layer, and processing is made from high to low (Dinedane 1995). The position of the NNW is represented each time in variable values, and the state of the NNW changes from one time to another when the input changes.

5.4.5 *Neural Networks as a Set of Relations*

In neural networks, units are organized into categories throughout a set of relations or synapses, where each synapse has a real value. This latter describes the impact of each unit on its neighbour, that is, if the impact is positive, the unit stimulates the other, but if the impact is negative, there will be no stimulation. Synapses will have one direction, that is, from input units to output units; however, if input units and output units are similar, synapses will have two directions (Tebelskis 1995).

5.4.6 *Programming*

It is the adaptation with the constraints of technology throughout a method. It is one field which establishes computer programmes that are similar to human intelligence and include the following: robot behaviour, language understanding and pattern recognition and knowledge representation. A programme, however, is a set of algorithms that are written with a language the computer can understand and execute (The Hutchinson Encyclopedia).

5.5 Acoustic Phonetics

It is a mistake to study phonology far from the acoustic side, that is, to rely only upon speech produced by the vocal organs, because acoustic data are found beforehand once trying to treat phonological units. It is through the ear that one can distinguish between sounds (De Saussure 2002, p. 51).

Acoustic phonetics is a branch which deals with the physical characteristics of speech sounds moving from the speaker's mouth to the listener's ear (Omar 1977, p. 22). It also studies the physics of the signal when the sounds travel in the air from the speaker to the listener because the speaker sends waves when he or she produces sounds that are transmitted in the form of vibrations in the air. These vibrations can be measured through mathematical techniques or through a spectrogram (Roach 1992, p. 8).

Acoustic phonetics also studies the relation between the activity of the oral cavity and sounds. It is interested in the material and the physical dimensions of the human speech during the transition phase. It studies height, pitch and amplitude, filtration and areas of resonance (Abdeljlil 1991, p. 301).

The means through which the air is transmitted is air; however there are other means such as glass and water. Air is considered as a mixture of gases which can be compressed or rarefied, and this means that the molecules can be gathered or separated; when organs of speech move, molecules found around the mouth move and therefore disturb other molecules which travel in the air (Connor 1973, p. 71).

Moving into the air, molecules can collide with each other, and some of them can rebound back to the starting point; therefore oscillation and vibration begin (Connor 1973).

The forms of air vibrations of speech sounds are very complicated. They can be categorized into two types: simple and complex. In addition, all the mechanisms of speech are the places of resonance (Abdeljlil 1991, p. 301).

Before embarking on the discussion of signal processing, it is worth mentioning some specific terms of this field.

5.5.1 Wave

In physics the movement of the wave can be defined as the mechanism through which energy is sent from one place to another into the air, spread out mechanically or propagated without the movement of matter (Birjandi and Salmani-Nodoushan 2005).

It is a set of alternated vibrations which are produced one after another. It is also defined as something that transmits oscillation or is a moveable oscillation. It is the spread out of the body forwards and backwards between a balanced position; whereas vibration is a set of oscillations (Fry 1979).

5.5.2 Wave Motion

The motion of the wave in physics is defined as the mechanism through which energy is transmitted from one place to another via waves that are spread out automatically without the motion of matter. It should be pointed out that in any place during the spread out of the wave, a periodic movement or an oscillation is found in a neutral position. Thus oscillation of molecules of air is similar to the sound transmitted into the air or water molecules, so in all these cases, the parts of the matter oscillate around its balanced position, and only the energy moves in a continuous way in one direction. These are called mechanical waves, because energy travels alone via a material medium without the mass movement of the medium itself (Birjandi and Salmani-Nodoushan 2005).

There are many types of waves: longitudinal waves, transverse waves and the combination between both. This is due to the transmission related to the motion itself.

(a) ***Longitudinal wave***: It is a wave in which the vibration is parallel to the direction of the motion. This wave is always mechanical because it is produced by successive compression and rarefication of the medium. Sound waves are of this type; in other words, sounds produced by the speaker generate a longitudinal wave consisting of air molecules that oscillate to reach the eardrum of the listener.
(b) ***Transverse wave***: It is a wave where vibrations are found in 90°, i.e. at right angles, with regard to the direction of the motion; the transverse wave can be mechanical like the wave, which appears in a stretched string.
(c) ***Combination between longitudinal and transverse wave***: It is a mechanical wave like waves found on the surface of water, and it results from a circular motion of water particles (Birjandi and Salmani-Nodoushan 2005).

5.5.3 Fundamentals and Harmonics

Most of the sounds that we hear and speech sounds are complex tones where a great deal of them is a mixture of frequencies; this mixture consists of a fundamental frequency and harmonics. The characteristics of sounds comprise the quality of frequencies found in this mixture and their average amplitudes.

Sounds are fundamental frequencies and a chain of harmonics. Harmonics is a group of the fundamental frequency; thus each complex tone is a mixture of frequencies if the fundamental is 100 Hz, and then the period is 10 m/s which denotes that the peak of the pressure is made every 10 m/s (Fry 1979).

5.5.4 Frequency

Frequency is the completed number of cycles in a second; thus vibration differs according to the different size of the body, its length, its form, its rate of intensity and its stretching (Issam 1992, p. 98).

5.5.5 Amplitude

It is the utmost movement far from the place of rest (Connor 1973, p. 73). It is also the distance within the movement of a vibrated body between the place of rest and the maximum movement. It can also be defined as the degree of air molecules movement within the wave which goes in accordance with the variations of air pressure of the wave. If the amplitude is greater, the power of the molecules that hits the eardrum will be very big; therefore the sound is loud (Fry 1979).

5.5.6 *Spectrum*

It is used in acoustic phonetics, and it is the segmentation of sounds into frequential components; this operation is similar to the segmentation of the white light into the colours of the rainbow; each sound consists of a different amount of energy in different frequencies (Fry 1979). It is also the visual representation of the acoustic signal; it displays the period of time and frequency of propagated energy in the signal. To get a spectrum, the fast Fourier transform is used; therefore degrees of amplitude appear in white and black colours. The white colour means there is no energy, while black colour means a large proportion of energy in many frequencies at the level of the vertical axis, whereas the period of time appears in the horizontal axis (Sean 2011, p. 80). The spectrum is also known as the identity card of the sound, i.e. it contains all the specific characteristics of the sound such as frequencies and amplitude (Fry 1979).

5.5.7 *Pitch*

Pitch can be defined as one characteristic of acoustic phonetics, and it is the name given to the frequency of the vibration of the vocal cords. In terms of music, it is a difference between two tones in terms of height and loudness though produced with two different musical instruments, such as piano and violin (Fry 1979).

5.5.8 *Sound*

It is the vibration of the air or any flexible medium where the vibration is transmitted through waves. When the speaker produces sounds, vibration is transmitted through waves with a high degree of speed, estimated at 340 m/s to the listener (Emerit 1977, p. 14). Speech sound is a set of overlapping waves of movement transmitted by carbon dioxide and formed in different configurations of speech organs (Birjandi and Salmani-Nodoushan 2005).

5.6 Signal Processing

Analytical tools are increasingly important to comprehend the process of linguistic communication. Speech analysis requires technical tools and techniques as well. These techniques can be summarized as follows (Lieberman and Blumstein 1988, p. 51):

- Conversion of continuous speech or analogue speech signal into a numerical representation
- Display of the stored data for selection
- Display of time and frequency

The field of signal analysis is related to acoustic phonetics, while digital signal processing belongs to the engineering sciences. It aims to get information from physical speech waves in addition to the development of methods of analogy. Scientists are concerned with the representation and transformation of the signals, which have a significant role in many applications, such as speech storage, speech synthesis and speech recognition (Birjandi and Salmani-Nodoushan 2005, p. 153).

5.6.1 Definition of Signal

Any physical element which is constant or variable in time is called a signal. There are two types of signals: permanent and transitional. The former stands for a signal which is continuous in time; the latter, however, grows stronger and dies away on the level of a horizontal axis (Neffati 1999, p. 12).

5.6.2 Sound Signal Variance

Speech recognition means to understand speech, that is, the listener recognizes a sentence or a word. However, the problem lies in the acoustic segments of the sentence or the word which are different, though produced by the same speaker on the one hand or from a group of speakers on the other, because speech starts as an analogue operation and ends as an acoustic signal continuous in time, in other words, the transformation of a variable stimulus to a constant response (Plomp 2002, p. 94).

5.6.3 Signal Processing Analysis

The goal of analysis is to provide a good and exact representation of the speech signal for an appropriate recognition. There are phases that the signal goes through during analysis: pretreatment and treatment in order to get a short-term spectrum. Speech recognition is possible thanks to the application of different data: acoustic phonetic, syntactic and semantic. The data fork of the aforementioned levels is a good demonstration for the complexity of speech recognition, in addition to the problem of variations found in the production of speech (Peinado and Segura 2006, p. 10).

Variations can be summarized as follows:

Intraspeaker variations due to characteristics of the speaker, interspeaker variations such as dialects, gender and noisy sounds, which can be influenced by the noise; therefore it makes the signal ambiguous. In this respect, the use of good tools such as neural networks and hidden Marcov models is necessary to cope with these variations (Peinado and Segura 2006).

5.6.4 Formants

A speech sound is considered air produced from a sound source which is converted by the vocal tract known as place of resonance. Vowels and some voiced sounds in particular rely on the vocal cords as a source, then sound outputs are filtered throughout resonators found in the vocal tract called formants. Some consonants are produced in some places of the vocal tract. When the air passes, it generates an aeroacoustic noise which is filtered by the oral cavity (Sean 2011, p. 07).

Frequencies of vowels are constant, while those of consonants are variable. There are five formants, which can be classified from low to high: F1, F2, F3, F4 and F5. Vowels in general contain more than three formants as regards the vowel **[æ];** the frequency of its first formant is estimated as **731 Hz**, while its second formant is estimated to **786 Hz**. As for the vowel [ɔ], F1 and F2 are more or less higher in frequency than those of the previous vowel **[æ]**. (F1 is 602 Hz, while F2 is 884 Hz.) The vowel **[i]** is different, i.e. F1 is **306 Hz**, while F2 is 2241 Hz. It is necessary to note that acoustic studies show that F1, F2 and F3 are necessary in determining the quality of the vowel. It should be noted that dimensions of vocal tracts are different for men, women and children, and therefore the formant frequencies of each group are totally different. (Plomp 2002, p. 94).

(a) ***Larynx***. It is a box which houses the vocal cords whose role is very important in producing the sounds. It is a strong box which consists of two cartilages: thyroid cartilage and cricoid cartilage. It is located immediately above the trachea and contains the vocal cords (they are similar to lips and are opposite the Adam's apple); they are attached to two small triangular cartilages called arytenoid cartilages (Roach 1991).

(b) ***Pharynx***. It is a passage similar to a pipe 8 cm long in men and 7 cm long in women, and it is connected to both the oral and nasal cavities. It amplifies the vibration coming from the vocal cords (Roach 1991).

5.6.4.1 Oral Cavity

It is the most important resonator in speech, because of the variability of its dimension and shape (Connor 1973, p. 34).

5.6.4.2 Nasal Cavity

It has a fixed dimension and shape; it is considered as the place of resonance. When the vocal cords vibrate, the lower jaw goes downward (Connor 1973).

5.7 Data Base

The number of sounds in Arabic is 34, 28 consonants and 6 vowels; 3 long vowels indicated in 3 consonants, ***waw***, ***ja*** and ***alif***; and 3 short vowels betokened in the diacritic marks ***fatha***, ***damma*** and ***kasra*** and the absence of the mark referred to as ***sukun***. The list of consonants starts with the /ʔ/ sound known as ***hamza*** and ends with the /j/ sound. To get a simple unit, one short vowel is added to a consonant, e.g. b+a; in contrast, to get a complex unit, many combinations are made, as shown below in types of syllables.

Short vowels, long vowels and minimal syllables are displayed in the following tables.

Arabic contains different types of syllables:

(a) ***Minimal syllable***. It consists of a consonant and a vowel. They are meaningful linguistic units, and they consist in prepositions, e.g. ***bi*** (by), ***fi***(in), ***li*** (for).
(b) ***Closed long syllable***. It consists of a consonant, a vowel and a consonant, e.g. *mithl* (like), *min* (from), *bal* (rather).
(c) ***Open long syllable***. It consists of a consonant and a long vowel, e.g. maa haa.
    Two forms related to the pause
(d) ***Long syllable closed with a consonant***. A consonant+a long vowel+a consonant, e.g. *kaan* (was).
(e) ***Long syllable closed with two consonants***. A consonant+a vowel+two consonants, e.g. *karb* and *fadl* (Kaddour 1999, p. 75).
(f) ***Final syllable known as tanwin***. A consonant plus a vowel plus the **[n]**, e.g. *man*, *mun*, and *mng*. Tables 5.1, 5.2, 5.3, 5.4, 5.5, 5.6, 5.7 show clearly Arabic long and short vowels, minimal syllables, open long syllables, closed syllables, syllables closed by a consonant, Final syllables closed by a consonant (Tanuin) and middle closed syllables.

Table 5.1 Arabic vowels

Number	Arabic written form	Transcription	Phonetic transcription
Short vowels			
1	ُ	***DAMMAH***	**u**
2	َ	***FATHAH***	**a**
3	ِ	***KASRAH***	**ɪ**
Long vowels			
1	و	***WAW***	**u:**
2	ا	***ALIF***	**a:**
3	ي	***YAE***	**i:**

Table 5.2 Minimal syllables

Number	Arabic written form	Latin written form	Phonetic transcription
1	ء	***a***	**ʔa**
2	ء	***u***	**ʔu**
3	ء	***ɪ***	**ʔɪ**

Table 5.3 Open long syllables

Number	Arabic written form	Latin written form	Phonetic transcription
1	آ	***HAMZAH***	**ʔa:**
2	أو	***HAMZAH***	**ʔu:**
3	إي	***HAMZAH***	**ʔi:**

Table 5.4 Closed syllables

Number	Arabic written form	Latin written form	Phonetic transcription
1	من	***MUN***	**mʊn**
2	من	***MAN***	**man**
3	من	***MIN***	**mɪn**

Table 5.5 Syllables closed by a consonant

Number	Arabic written form	Latin written form	Phonetic transcription
1	مان	***MAAN***	**ma :n**
2	مون	***MUUN***	**mu :n**
3	مين	***MIIN***	**mi :n**

Table 5.6 Final syllables closed by a consonant (Tanuin)

Number	Arabic written form	Latin written form	Phonetic transcription
1	بن	***BUN***	**bʊn**
2	بن	***BAN***	**Ban**
3	بن	***BIN***	**bɪn**

Table 5.7 Middle closed syllables

Number	Arabic written form	Latin written form	Phonetic transcription
1	أي	**AY**	**aj**
2	أو	**AW**	**aw**

5.7.1 Knowledge Base

The use of sound combinations will result in obtaining some unacceptable sequence of sounds with regard to the Arabic phonological and morphological system. In this respect, deletion of some sequences is done. Application of factorial analysis is very important to eliminate the repeated and some compound sounds. The use of factorial analysis requires showing the word formation to be able to eliminate some sounds because sound elimination is done on the basis of opposition or repetition.

Arabic word formation is triliteral, quadrilateral and quintuple. The trilateral word can be shown in the verb ***/faʕala/*** (فعل) and represented in **1, 2, 3** since it contains three consonants; therefore **1** and **2** and **1**, **2**, and **3** are never repeated. Quadrilateral may be displayed in ***/fa:ʕɪlun/*** (فاعل) and can be represented **1,2,3,4**. It should be noted that **1,2** cannot be repeated, **1.2.3** cannot be repeated either and **1,3,4** are the like. Quintuple literal is shown in ***/mafa:ʕɪlun/*****1,2,3,4,5,** where **1,2,3** are not repeated and **1,3,5** are not repeated either; however **1,3,4,5** are repeated.

5.7.2 Deletion by Opposition

As has been mentioned before, this work tries to study the Arabic sounds where their combination results in complex sounds that are rejected by the classical Arabic. The sounds are ts, dʒ, ʒr,zʒ and tr (Tables 5.8 and 5.9).

Table 5.8 Word formation

Acceptable redundancy	Unacceptable redundancy	Unacceptable redundancy	System	Triliteral root
1,3	**1,2,3**	**1,2**	**1,2,3**	
	1-2-3-4	**1,2**	**1,2,3,4**	**Quadrilateral root**
1-2-3-4-5	**1,2,3**	**1,3,4**	**1,2,3,4,5**	

Table 5.9 Elimination by opposition

Sound	Compound sound	Elimination	Foreign sound
ت	تس ts	ت **t**	س**s**
د	دج dʒ	د **d**	ج **ʒ**
ج	جر dr	ج **ʒ**	ر**r**
ز	زج ʒz	ز **z**	ج **ʒ**
ت	تر tr	ت **t**	ر**r**

5.7.3 Stages of Combination

The combination of sounds contains six stages; each stage consists of three operations except for the last one which contains only two stages because it deals with mid syllables; therefore it requires only two semivowels.

As mentioned before, Arabic has 28 consonants and 6 vowels: 3 long and 3 short.

NB: Arabic consonants start with **/ ʔ/** which is represented in (ء) and end in /j/. Each phase goes through three operations, except for the sixth one which has only two operations.

First Phase

First operation. There are three vowels, three long and three short, ***fathah***, ***dammah*** and ***kasrah***, so the first consonant is multiplied by 28 consonants, i.e. *ʔa* is multiplied to all consonants, e.g. ***ʔa* x *ʔa*, *ʔa* x *ba*, *ʔ a* x *ta*, *ʔa* x *tha*, *ʔa* x *ha***, etc.

Second operation. The first consonant ***ʔ*** and /u/, i.e. a minimal syllable, is multiplied to all consonants, that is, ***ʔu*** is multiplied to the 28 consonants, e.g. ***ʔu* x *ba*, *ʔu* x *ta*, *ʔu* x *tha*, *ʔu* x *ha***, etc.

Third operation. The consonants ***ʔ*** and **i**; again a minimal syllable is multiplied by all the 28 consonants, e.g. ***ʔi* x *ba*, *ʔi* x *ta*, *ʔi* x *tha*, *ʔi* x *ha***, etc.

Second Phase

This step contains long units, that is, long i, long u and long a, which are, then, mixed with ***/ʔ/*** and give ***ʔa:***, ***ʔu:***, and ***ʔi:*** (آ، أو ، إيـ))

First operation. This unit ***/ʔa:/*** is multiplied by the 28 consonants, e.g. ***ʔa:* x *ba*, *ʔa:* x *ta*, *ʔa:* x *tha*, *ʔa:* x *ha***, etc.

Second operation. This unit ***/ʔu:/*** is multiplied by the 28 consonants, e.g. ***ʔu:* x *ba*, *ʔu:* x *ta*, *ʔu:* x *tha*, *ʔu:* x *ha***, etc.

Third operation. The following unit ***/ʔi:/*** is multiplied by the 28 consonants, e.g. ***ʔi:*x *ba*, *ʔi:*x *ta*, *ʔi:* x *tha*, *ʔi:* x *ha***, etc.

Third Phase

This phase contains three closed syllables containing a consonant, a vowel and a consonant, e.g. ***ʔan***, ***ʔun*** and ***ʔin***.

First operation. The following unit ***/ʔan/*** is multiplied by the 28 consonants of the data base, that is to say, ***ʔan* x *ʔa*, *ʔan* x *ba*, *ʔan* x *ta*, *ʔan* x *tha*** and ***ʔan* x *ha*** until the last consonant. Then the next syllable that is formed with the ***b*** consonant is dealt with, i.e. ***ban* x *ʔa*, *ban* x *ba*, *ban* x *ta*, *ban* x *tha*** and ***ban* x *ha***, until the last letter of the data base. The same process is made with the next letter until the last one, i.e. ***tan* x *ʔa*, *tan* x *ba*, *tan* x *ta*, *tan* x *tha*** and ***tan* x *ha***, and until the last consonant the same step is followed.

Second operation. This operation consists in the following: */?un/* is multiplied by all consonants of the data base, i.e. ***?un*** x ***?a***, ***?un*** x ***ba***, ***?un*** x **ta**, ***?un*** x ***tha***, and ***?un*** x ***ha***, etc., until the last consonant, then we move to the second syllable, i.e. ***bun*** x ***?a***, ***bun*** x ***ba***, ***bun*** x ***ta***, ***bun*** x ***tha*** and ***bun*** x ***ha***, until the last consonant of the data base. The same step is followed until the last consonant.

Third operation. The third operation in this phase starts as follows:

The syllable */?in/* is multiplied by all consonants of the data base, i.e. ***?in*** x ***?a***, ***?in*** x ***ba***, ***?in*** x ***ta***, ***?in*** x ***tha***, ***?in*** x ***?a***, ***?in*** x ***ba***, ***?in*** x ***ta*** ***?in*** x ***tha***, ***?in*** x ***ha***, etc., then ***bin*** x ***?a***, ***bin*** x ***ba***, ***bin*** x ***ta***, ***bin*** x ***tha***, ***bin*** x ***ha***, etc. until the last consonants of the data base, then the third syllable */tin/* is multiplied by the previous consonants as done with the previous syllable, that is, ***tin*** x ***?a***, ***tin*** x ***ba***, ***tin*** x ***ta***, ***tin*** x ***tha***, and ***tin*** x ***ha*** until the last consonant.

Fourth Phase

This phase deals with syllables containing a consonant, a long vowel and a consonant, e.g. ***?a:n***, ***? u:n*** and ***?i:n***. The generation of sounds in this phase is the other way round.

First operation. */?/* with the absence of the mark referred to as ***sukun*** is multiplied to ***?a:n***, that is, ***?*** x ***?a:n***, then **b**x ***?a:n***, then ***t*** x ***?a:n***, then ***th*** x ***?a:n***, then ***h***x ***?a:n***, etc. until the last consonant in the list. Next the same sound */ ?/* is multiplied to ***ba:n***, ***b*** is multiplied to ***ba:n***, then ***t*** is multiplied to ***ba:n***, then ***th*** is multiplied to ***ba:n***, then ***h*** is multiplied to ***ba:n***, etc. until the last consonant. Afterwards we move to the second syllable with the sound ***b***, that is, ***?*** x ***ba:n***, ***b***x ***ba:n***, ***t*** x ***ba:n***, ***th*** x ***ba:n***, ***h*** x ***ba:n***, etc. until the last consonant in the list.

The third syllable with the sound */t/*, that is, ***?*** x ***t a:n***, ***b***x ***t a:n***, ***t*** x ***t a:n***, ***th*** x ***ta:n***, ***h***x ***ta:n***, etc. until the last consonant in the list. The same step is done, i.e. each time we move to another consonant we add the unit ***a:n***.

Second operation. This operation consists in the following: the consonant */?/* with the absence of the vowel is multiplied to the syllable ***?u:n***, in other words, ***?*** x ***?u:n***, ***b***x ***?u:n***, ***t*** x ***?u:n***, ***th*** x ***?u:n***, ***h***x ***?u:n***, etc. until the last consonant in the list, then we move to the next syllable containing the *b* sound, i.e. ***?*** x ***b u:n***, ***b***x ***bu:n***, ***t*** x ***b u:n***, ***th*** x ***bu:n***, ***h***x ***bu:n***, etc. until the last consonant in the list, then the next syllable that contains the consonant ***t*** is dealt with, that is, ***?*** x ***tu:n***, ***b***x ***tu:n***, ***t*** x ***tu:n***, ***th*** x ***tu:n***, ***h***x ***tu:n***, etc. until the last consonant in the list. The same step is followed, i.e. each time we move to a consonant, the unit ***u:n*** is added.

Third operation. This operation deals with the consonant ***?*** without the addition of a vowel, that is, ***?*** x ***? i:n***, ***b***x ***? i:n***, ***t*** x ***? i:n***, ***th*** x ***? i:n***, ***h***x ***?i:n***, etc. until the last consonant in the list. The next syllable comes afterwards, that is, the syllable that entails the sound ***t*** is multiplied to all the consonants of the list, i.e. ***?*** x ***ti:n***, ***b***x ***ti:n***, ***t*** x ***ti:n***, ***th*** x ***ti:n***, ***h***x ***ti:n***, etc. the same process is done, that is, each time we pass to another consonant, we add the unit ***i:n***.

Fifth Phase

In this phase there are syllables which contain a consonant, a vowel and a consonant and are final syllables, e.g. ***ʔan***, ***ʔun*** and ***ʔin***. In this phase we do the reverse operation, i.e. /ʔ/ is multiplied to the aforementioned syllables.

First operation. /***ʔ***/ alone without the addition of any vowel is multiplied to all the syllables of the data base. The process is made as follows: ***ʔ* x *ʔ an*, *b*x *ʔ an*, *t* x *ʔ an*, *th* x *ʔ an*, *h*x *ʔ an***, etc. until the last consonant of the list, then we move to the second syllable, which is made of the consonant *b*, then is multiplied by all the sounds of the list, that is, ***ʔ* x *ban*, *b*x *ban*, *t* x *ban*, *th* x *ban*, *h*x *ban***, etc.; the third syllable is multiplied by all the sounds of the list afterwards, i.e. ***ʔ* x *t an*, *b*x *t an*, *t* x *tan*, *th* x *tan*, *h*x *tan***, etc., the same process is done, i.e. the next syllable is formed with the next consonant and is multiplied by all the consonants of the list.

Second operation. This operation is very similar to the previous one in terms of combination; however it is different in the type of syllable, because the previous syllable comprises the vowel [**a**], while this syllable contains the vowel [**u**]. It may be explained as follows: ***ʔ*x *ʔun*, *b*x *ʔun*, *t* x *ʔun*, *th* x *ʔun*, *h*x *ʔun***, etc., until the last consonant of the list, then we move to the second syllable which is made by the sound /b/ and multiplied it by all the sounds of the list, that is, **ʔx *bun*, *b*x *bun*, *t* x *bun*, *th* x *bun*, *h*x *bun***, etc., the third syllable is multiplied by all the sounds of the list afterwards, i.e. **ʔ x *t un*, *b*x *t an*, *t* x *tun*, *th* x *tun*, *h*x *tun***, etc. The same process is done, i.e. the next syllable is formed with the next consonant and is multiplied to all the consonants of the list.

Third operation. Again this operation is very similar to the first and second previous operations in terms of combination but different only in terms of syllable; in other words, instead of **/un/** and **/an/**, which are used to form syllables, **/in/** is used to form the syllable. The process is done as follows: ***ʔ* x*ʔ in*, *b*x *ʔin*, *t* x *ʔin*, *th* x *ʔin*, *h*x *ʔin***, etc. until the last consonant in the list, then ***ʔ*x *b in*, *b*x *bin*, *t* x *bin*, *th* x *bin*, *h*x *bin*, *etc***. and then ***ʔ*x *tin*, *b*x *tin*, *t* x *tin*, *th* x *tin*, *h*x *tin***, etc., the same step is followed with all consonants.

Sixth Phase

This phase contains only two operations because it works only with two sounds, /w/ and /j/, which are used to form intermediary syllables.

First operation. The simple units are multiplied by **w** in the following way: ***ʔa* x *w*, *ba* x *w*, *ta* x *w***, and ***tha* x *w*** until the last consonant.

Second operation. Simple units are multiplied by j, e.g. ***ʔa* x *y*, *ba* x *y*, *ta* x *y*,** and ***tha* x *y*** until the last consonant, i.e. ***ya* x *y***.

	1	2	3	4	5	6	7	8	9	10
ʔa,ʔu,ʔi	ʔaxʔa	baxʔa	taxʔa	θaxʔa	ʒaxʔa	haxʔa	xaxʔa	daxʔa	ðaxʔa	raxʔa
	ʔaxba	Baxba	Taxba	Θaxba	ʒaxba	Haxba	Xaxba	daxba	Ɖaxba	raxba
	ʔaxta	Baxta	Taxta	Θaxta	ʒaxta	Haxta	Xaxta	daxta	Ɖaxta	Raxta
ba,bu,bi	ʔaxθa	Baxθa	Taxθa	θaxθa	ʒaxθa	Haxθa	Xaxθa	daxθa	Ɖaxθa	Raxθa
	ʔaxʒa	baxʒa	taxʒa	θaxʒa	ʒaxʒa	haxʒa	xaxʒa	daxʒa	ðaxʒa	raxʒa
	ʔaxha	Baxha	Taxha	Θaxha	ʒaxha	Haxha	Xaxha	daxha	Ɖaxha	raxha
ta,tu,ti ti	ʔaxxa	Baxxa	Taxxa	Θaxxa	ʒaxxa	Haxxa	Xaxxa	daxxa	Ɖaxxa	raxxa
	ʔaxda	Baxda	Taxda	Θaxda	ʒaxda	Haxda	Xaxda	daxda	Ɖaxda	raxda
	ʔaxða	Baxða	Taxða	Θaxða	ʒaxða	Haxða	Xaxða	daxða	ðaxða	raxða
Θa,θu,θi	ʔaxra	Baxra	Taxra	Θaxra	ʒaxra	Haxra	Xaxra	daxra	čaxra	Raxra
	ʔaxza	Baxza	Taxza	Θaxza	ʒaxza	Haxza	Xaxza	daxza	čaxza	Raxza
	ʔaxsa	Baxsa	Taxsa	Θaxsa	ʒaxsa	Haxsa	Xaxsa	daxsa	čaxsa	Raxsa
ʒa, ʒu,ʒi	ʔaxʃa	baxʃa	taxʃa	θaxʃa	ʒaxʃa	haxʃa	xaxʃa	daxʃa	čaxʃa	raxʃa
	ʔaxʂa	Baxʂa	Taxʂa	Θaxʂa	ʒaxʂa	Haxʂa	Xaxʂa	daxʂa	ðaxʂa	Raxʂa
	ʔaxḍa	baxḍa	taxḍa	θaxḍa	ʒaxḍa	haxḍa	xaxḍa	daxḍa	ðax ḍa	rax ḍa
ha,hu,hi	ʔaxṭa	baxṭa	taxṭa	θaxṭa	ʒaxṭa	haxṭa	xaxṭa	daxṭa	ðax ṭa	rax ṭa
	ʔaxða	baxða	taxða	θaxða	ʒaxða	haxða	xaxða	daxða	ðax ða	rax ða
	ʔaxʕa	baxʕa	taxʕa	θaxʕa	ʒaxʕa	haxʕa	xaxʕa	daxʕa	ðax ʕa	raxʕa
xa,xu,xi	ʔaxɤa	baxɤa	taxɤa	θaxɤa	ʒaxɤa	haxɤa	xaxɤa	daxɤa	ðax ɤa	raxɤa
	ʔaxfa	Baxfa	Taxfa	Θaxfa	ʒaxfa	Haxfa	Xaxfa	daxfa	ðaxfa	Raxfa
	ʔaxqa	Baxqa	Taxqa	Θaxqa	ʒaxqa	Haxqa	Xaxqa	axqa	ðaxqa	raxqa
da,du,di	ʔaxka	Baxka	Taxka	Θaxka	ʒaxka	Haxka	Xaxka	daxka	ðaxka	raxka
	ʔaxla	Baxla	Taxla	Θaxla	ʒaxla	haxla	xax la	daxla	ðaxla	Raxla
	ʔaxma	Baxma	Taxma	Θaxma	ʒaxma	Haxma	Xaxma	daxma	ðaxma	raxma

ða,ðu,ði	ʡaxna	Baxna	Taxna	Θaxna	ʒaxna	haxna	xax na	daxna	ðaxna	raxna
	ʡaxha	Baxha	Taxha	Θaxha	ʒaxha	haxha	xaxḥ a	daxha	ðaxḥ a	rax ḥ a
	ʡaxwa	Baxwa	Taxwa	Θaxwa	ʒaxwa	haxwa	Xaxwa	daxwa	ðaxwa	raxwa
ra,ru,ri,ri	ʡaxja	Baxja	Taxja	Θaxja	ʒaxja	haxja	xaxja	daxja	ðax ja	Raxja
	First operation									
	11	**12**	**13**	**14**	**15**	**16**	**17**	**18**	**19**	**20**
za,zu,zi,zi	zaxʡa	saxʡa	ʃaxʡa	şaxʡa	d̜axʡa	t̜axʡa	ðaxʡa	ʢaxʡa	ɤaxʡa	faxʡa
	Zaxba	Saxba	ʃaxba	Şaxba	d̜axba	t̜axba	ðaxba	ʢaxba	ɤaxba	faxba
	Zaxta	Saxta	ʃaxta	Şaxta	d̜axta	t̜axta	ðaxta	ʢaxta	ɤaxta	Faxta
sa,su,siiii	ZaxΘa	SaxΘa	ʃaxΘa	ŞaxΘa	d̜axΘa	t̜axΘa	ðaxΘa	ʢaxΘa	ɤa xΘa	faxΘa
	zaxʒa	saxʒa	ʃaxʒa	şaxʒa	d̜axʒa	t̜axʒa	ðaxʒa	ʢaxʒa	ɤaxʒa	faxʒa
	Zaxha	Saxha	ʃa xha	şa xha	d̜a xha	t̜a xha	ða xha	ʢa xha	ɤa xha	fa xha
ʃa,ʃu,ʃi	Zaxxa	Saxxa	ʃax xa	şax xa	d̜ax xa	t̜ax xa	ðax xa	ʢax xa	ɤax xa	fax xa
	Zaxda	Saxda	ʃaxda	Şaxda	d̜axda	t̜axda	ðaxda	ʢaxda	ɤaxda	faxda
	Zaxða	Saxða	ʃax ða	şax ða	d̜ax ða	t̜ax ða	ðax ða	ʢax ða	ɤax ða	fax ða
sa,şu,şi	Zaxra	Saxra	ʃaxra	Şaxra	d̜axra	t̜axra	ðaxra	ʢaxra	ɤaxra	Faxra
	Zaxza	Saxza	ʃaxza	Şaxza	d̜axza	t̜axza	ðaxza	ʢaxza	ɤaxza	Faxza
	Zaxsa	Saxsa	ʃaxsa	Şaxsa	d̜axsa	t̜axsa	ðaxsa	ʢaxsa	ɤaxsa	Faxsa
d̜a,d̜u,d̜i	zaxʃa	saxʃa	ʃaxʃa	şaxʃa	d̜axʃa	t̜axʃa	ðaxʃa	ʢaxʃa	ɤaxʃa	faxʃa
	Zaxşa	Saxşa	ʃaxşa	Şaxşa	d̜axşa	t̜axşa	ðaxşa	ʢaxşa	ɤaxşa	Faxşa

It should be pointed out that some samples regarding the simple unit, that is, the minimal syllables (ء) which is equivalent to /ʔ/ are provided to clarify the process of combination, and then a rule is given to eliminate sequences that are rejected by the Arabic system.

5.7.3.1 Combination of Minimal Syllable

The tables mentioned above represent the combination of simple units, that is, the minimal syllables containing a consonant and a vowel; the vowel may be either an /a/, an /u/ or an /i/. Combination can result in having some inacceptable syllables with regard to the Arabic system and should be eliminated.

The rule is as follows:

If cv1+cv2 are the same, then they are deleted.

If cv1+cv2 are not the same and do not contribute to form a meaning, then they are deleted.

	Zaxsa	Saxsa	ʃaxsa	Şaxsa	ḑaxsa	ţaxsa	ðaxsa	ʕaxsa	ɤaxsa	faxsa
ḑa,ḑu,ḑi	zaxʃa	saxʃa	ʃaxʃa	şaxʃa	ḑaxʃa	ţaxʃa	ðaxʃa	ʕaxʃa	ɤaxʃa	faxʃa
	Zaxşa	Saxşa	ʃaxşa	Şaxşa	ḑaxşa	ţaxşa	ðaxşa	ʕaxşa	ɤaxşa	Faxşa
	zaxḑa	saxḑa	ʃaxḑa	şaxḑa	ḑaxḑa	ţaxḑa	ðaxḑa	ʕaxḑa	ɤax ḑa	fax ḑa
ţa, ţu, ţi	zaxţa	zaxţa	axţa	şaxţa	ḑaxţa	ţaxţa	ðaxţa	ʕaxţa	ɤax ţa	fax ţa
	zax ða	sax ða	ʃax ða	şax ða	ḑaxða	ţaxða	ðaxða	ʕax ða	ɤax ða	fax ða
	zax ʕa	zax ʕa	ʃax ʕa	şax ʕa	ḑaxʕa	ţaxʕa	ðaxʕa	ʕax ʕa	ɤax ʕa	fax ʕa
ða,ðu,ði	zax ɤa	sax ɤa	ʃax ɤa	şax ɤa	ḑaxɤa	ţaxɤa	ðaxɤa	ʕax ɤa	ɤax ɤa	fax ɤa
	zax fa	sax fa	ʃax fa	şax fa	ḑaxfa	ţaxfa	ðaxfa	ʕax fa	ɤax fa	fax fa
	zax qa	sax qa	ʃax qa	şax qa	ḑaxqa	ţaxqa	ðaxqa	ʕax qa	ɤax qa	fax qa
ʕa,ʕu,ʕi	zax ka	sax ka	ʃax ka	şax ka	ḑaxka	ḑaxka	ðaxka	ʕax ka	ɤax ka	fax ka
	zax la	sax la	ʃax la	şax la	ḑax la	ḑax la	ðaxla	ʕax la	ɤax la	fax la
	zax ma	sax ma	ʃax ma	şax ma	ḑaxma	ḑaxma	ðaxma	ʕaxma	ɤaxma	Faxma
ɤa,ɤu,ɤi	zax na	sax na	ʃax na	şax na	ḑax na	ḑax na	ðax na	ʕax na	ɤax na	fax na
	zax ḥ a	sax ḥ a	ʃax ḥ a	şax ḥ a	ḑax ḥ a	ḑax ḥ a	ðax ḥ a	ʕax ḥ a	ɤax ḥ a	fax ḥ a
	zaxwa	Saxwa	ʃaxwa	şax wa	ḑaxwa	ḑaxwa	ðaxwa	ʕaxwa	ɤaxwa	faxwa
fa, fu, fi	zaxja	Saxja	ʃax ja	şax ja	ḑax ja	ḑax ja	ðax ja	ʕaxja	ɤaxja	fax ja

	First operation							
Number	**21**	**22**	**32**	**24**	**25**	**26**	**27**	**28**
	qaxʔa	kaxʔa	laxʔa	maxʔa	naxʔa	haxʔa	waxʔa	jaxʔa
qa,qu,qi	qaxba	Kaxba	Laxba	Maxba	Naxba	haxba	Waxba	Jaxba
	qaxta	Kaxta	Laxta	Maxta	Naxta	haxta	Waxta	Jaxta
	qaxθa	Kaxθa	Laxθa	Maxθa	Naxθa	haxθa	waxθa	Jaxθa
ka, ku, ki	qaxʒa	kaxʒa	laxʒa	maxʒa	naxʒa	haxʒa	waxʒa	jaxʒa
	qaxha	Kaxha	lax xha	Maxha	Naxha	haxha	waxha	Jaxha

	qaxxa	Kaxxa	Laxxa	Maxxa	Naxxa	haxxaa	waxxa	Jaxxa
la,lu, li	qaxda	Kaxda	Laxda	Maxda	Maxda	haxda	Waxda	Jaxda
	qaxða	kax ða	lax ða	max ða	nax ða	haxðaa	wax ða	jaxða
	qaxra	Kaxra	Laxra	Maxra	Naxra	ḥ axra	Waxra	jaxra
ma,mu,mi	qaxza	Kaxza	Laxza	Maxza	Naxza	ḥ axza	Waxza	jaxza
	qaxsa	Kaxsa	Laxsa	Maxsa	Naxsa	ḥ axsa	Waxsa	jaxsa
	qaxʃa	kaxʃa	laxʃa	maxʃa	naxʃa	ḥ axʃa	waxʃa	jaxʃa
na, nu, ni	qaxşa	Kaxşa	Laxşa	Maxşa	Naxşa	ḥ axşa	Waxşa	jaxşa
	qaxḑa	kax ḑa	lax ḑa	max ḑa	nax ḑa	ḥ ax ḑa	wax ḑa	jaxḑa
	qaxţa	kax ţa	lax ţa	max ţa	nax ţa	ḥ ax ţa	wax ţa	jax ţa
ḥ a,ḥ u,ḥ i	qaxða	kax ða	lax ða	max ða	nax ða	ḥ ax ða	wax ða	jax ða
	qaxʕa	kax ʕa	lax ʕa	max ʕa	nax ʕa	ḥ ax ʕa	wax ʕa	jax ʕa
	qaxɤa	kax ɤa	lax ɤa	max ɤa	nax ɤa	ḥ ax ɤa	wax ɤa	jax ɤa
wwa,wu,wu,wi	qaxfa	kax fa	lax fa	max fa	nax fa	ḥ ax fa	wax fa	jax fa
	qaxqa	kax qa	lax qa	max qa	nax qa	ḥ ax qa	wax qa	jax qa
	qaxka	kax ka	kax ka	max ka	nax ka	ḥ ax ka	wax ka	jax ka
Ja, Ju, Ji	qaxla	kax la	lax la	max la	nax la	ḥ ax la	wax la	jax la
	qaxma	kax ma	lax ma	max ma	nax ma	ḥ ax ma	wax ma	jax ma
	qaxna	kax na	lax na	max na	nax na	ḥ ax na	wax na	jax na
	qaxha	kax ḥ a	lax ḥ a	max ḥ a	nax ḥ a	ḥ ax ḥ a	wax ḥ a	jax ḥ a
	qaxwa	kax wa	lax wa	max wa	nax wa	ḥ ax wa	wax wa	jax wa
	qaxja	kax ja	lax ja	max ja	nax ja	ḥ ax ja	wax ja	wax ja

As for complex units which contain a consonant, a long vowel, /n/ sound, a consonant and a vowel, the rule is if cvv+c(n) +cv do not form a meaning, then they are deleted.

5.8 Syllables Recognition Using Neural Networks

The selected words for applying automatic recognition are three: *ʔamara*, *qaraʔa* and *θaʔara*, where the syllable / *ʔa*/ is found in first position, mid position and last position. The table mentioned below displays the words and the location of the syllables.

Word	First syllable	Medial syllable	Final syllable
أمر	أ		
ثأر		أ	
قرأ			أ

Before applying the process of automatic recognition, it is worth mentioning that the use of symbols to codify data that can be comprehended by the machine using neural networks is of a great importance. The cipher of this data is typified in letters and numbers. The letters used for coding are A, h, d, m, g and f, where A symbolizes the minimal syllable known as *hamza* in Arabic and a short vowel, **h** represents the word (homme), **d** stands for (debut), i.e. onset, **m** denotes the medial syllable and **g** indicates garcons.

As far as the letter **f** is concerned, there are two: one means final, and the other means femme (woman). As for the numbers, they represent speakers. The words mentioned above were recorded by 20 informants of both genders and different ages, 5 men, 5 women, 5 boys and 5 girls, then segmented to have the (**ء**) referred to as ***hamza*** and is equivalent to ***/ʔa/*** which is located in first, mid, minimal syllable and final position.

Two charts are used for the process of recognition; one represents the referential sample and the other the test sample. We mean by referential sample input and output of sounds. Inputs are numerical values of formants, amplitude and period, while outputs are the recorded sounds**.** Training neural networks to recognize sounds requires three things: frequency of formants, amplitude and period which are put in the upper layer as nodes and then connected to a hidden layer which in turn is connected to a lower layer that represents outputs. The tables below show the characteristics: formants, intensity and period of the minimal syllable ***/ʔa/*** located in first, medial and final position after being produced by each speaker, then recorded and treated automatically. Table 5.10 represents the two charts.

Table 5.10 Numerical values of referential samples

Referential samples						
Sound	F1	F2	F3	F4	Int	Per
Ahd1	**658 Hz-14 dB**	**1076 Hz-18 dB**	**2668 Hz-3 dB**	**X**	**0.1566 dB**	**0.0001667**
Ahm1	**641 Hz-18 dB**	**1037 Hz-20 dB**	**1692 Hz-1 dB**	**2504 Hz-0 dB**	**0.2608 dB**	**0.0001667**
Ahf1	**580 Hz-18 dB**	**928 Hz-16 dB**	**1621 Hz-3 dB**	**2633 Hz-4 dB**	**0.2602 dB**	**0.0001667**
Ahd2	**698 Hz-11 dB**	**1082 Hz-2 dB**	**2152 Hz-3 dB**	**X**	**0.0966 dB**	**0.0001667**
Ahm2	**721 Hz-17 dB**	**1251 Hz-9 dB**	**2555 Hz-7 dB**	**X**	**0.1283 dB**	**0.0001667**
Ahf2	**652 Hz-20 dB**	**992 Hz-20 dB**	**2618Hz-8 dB**	**X**	**0.1623 dB**	**0.0001667**
Ahd3	**670Hz-17 dB**	**1133Hz-14 dB**	**2167Hz-5 dB**	**X**	**0.1609 dB**	**0.0001667**
Ahm3	**692Hz-19 dB**	**1337Hz-14 dB**	**2393Hz-6 dB**	**X**	**0.1294 dB**	**0.0001667**
Ahf3	**708 Hz-17 dB**	**1056 Hz-14.5 dB**	**2528 Hz-10 dB**	**X**	**0.1956 dB**	**0.0001667**
Ahd4	**659 Hz-15 dB**	**1154 Hz-17 dB**	**2445 Hz-9 dB**	**X**	**0.1632 dB**	**0.0001667**
Ahm4	**625 Hz-23 dB**	**1025 Hz**	**1649 Hz-8 dB**	**2489 Hz-6 dB**	**0.1912 dB**	**0.0001667**
Ahf4	**640 Hz-8 dB**	**1069 Hz-17 dB**	**2660 Hz-2 dB**	**X**	**0.1635 dB**	**0.0001667**
Ahd5	**670 Hz-16 dB**	**1140 Hz-16 dB**	**2714 Hz-8 dB**	**X**	**0.1306 dB**	**0.0001667**
Ahm5	**698 Hz-16 dB**	**1204 Hz-13 dB**	**2640 Hz-3 dB**	**X**	**0.1632 dB**	**0.0001667**
Ahf5	**652 Hz-17 dB**	**1014 Hz-16 dB**	**2696 Hz-4 dB**	**X**	**0.1628 dB**	**0.0001667**
Afd1	**775 Hz-16 dB**	**1182 Hz-17 dB**	**2152 Hz-0 dB**	**X**	**0.1840 dB**	**0.0001667**
Afm1	**768 Hz-16 dB**	**1202 Hz-12 dB**	**2231 Hz-2 dB**	**X**	**0.0949 dB**	**0.0001667**
Aff1	**700 Hz-18 dB**	**1115 Hz-20 dB**	**2144 Hz-0 dB**	**X**	**0.2194 dB**	**0.0001667**
Afd2	**803 Hz-14 dB**	**999 Hz-6 dB**	**1649 Hz-10 dB**	**2541 Hz-5 dB**	**0.1245 dB**	**0.0001667**
Afm2	**823 Hz-13 dB**	**1087 Hz-11 dB**	**1536 Hz-12 dB**	**2549 Hz-4 dB**	**0.1251 dB**	**0.0001667**
Aff2	**898 Hz**	**1387 Hz**	**2340 Hz**	**X**	**0.1274 dB**	**0.0001667**
Afd3	**792 Hz**	**1132 Hz**	**1534 Hz**	**X**	**0.0972 dB**	**0.0001667**
Afm3	**896 Hz**	**1525 Hz**	**X**	**X**	**0.0972 dB**	**0.0001667**

(continued)

Table 5.10 (continued)

Referential samples						
Sound	F1	F2	F3	F4	Int	Per
Aff3	**835 Hz**	**1399 Hz**	**X**	**X**	**0.1603 dB**	**0.0001667**
Afd4	**717 Hz**	**1058 Hz**	**1454 Hz**	**2520 Hz**	**0.2603 dB**	**0.0001667**
Afm4	**835 Hz**	**1020 Hz**	**1510 Hz**	**2644 Hz**	**0.1312 dB**	**0.0001667**
Aff4	**796 Hz**	**1065 Hz**	**1339 Hz**	**2645 Hz**	**0.1953 dB**	**0.0001667**
Afd5	**875 Hz**	**1462 Hz**	**2625 Hz**	**X**	**0.1962 dB**	**0.0001667**
Afm5	**813 Hz**	**1002 Hz**	**1520 Hz**	**2638 Hz**	**0.1286 dB**	**0.0001667**
Aff5	**871 Hz**	**1419 Hz**	**2571 Hz**	**X**	**0.2577 dB**	**0.0001667**
Agd1	**683 Hz**	**1038 Hz**	**1359 Hz**	**2269 Hz**	**0.1600 dB**	**0.0001667**
Agm1	**721 Hz**	**1165 Hz**	**2480 Hz**	**X**	**0.1603 dB**	**0.0001667**
Agf1	**801 Hz**	**1098 Hz**	**1986 Hz**	**X**	**0.2899 dB**	**0.0001667**
Agd2	**911 Hz**	**1257 Hz**	**2225 Hz**	**X**	**0.1563 dB**	**0.0001667**
Agm2	**564 Hz**	**1550 Hz**	**2520 Hz**	**X**	**0.1872 dB**	**0.0001667**
Agf2	**646 Hz**	**918 Hz**	**1255 Hz**	**2180 Hz**	**0.2214 dB**	**0.0001667**
Agd3	**537 Hz**	**1127 Hz**	**1654 Hz**	**2051 Hz**	**0.1878 dB**	**0.0001667**
Agm3	**541 Hz**	**1032 Hz**	**1658hz**	**1818 Hz**	**0.1560 dB**	**0.0001667**
Agf3	**653 Hz**	**983 Hz**	**1352 Hz**	**2299 Hz**	**0.2275 dB**	**0.0001667**
Agd4	**853 Hz**	**1096 Hz**	**1584 Hz**	**2411 Hz**	**0.1953 dB**	**0.0001667**
Agm4	**763 Hz**	**1243 Hz**	**1635 Hz**	**2528 Hz**	**0.1291 dB**	**0.0001667**
Agf4	**785 Hz**	**1215 Hz**	**1782 Hz**	**2416 Hz**	**0.2539 dB**	**0.0001667**
Agd5	**688 Hz**	**972 Hz**	**1347 Hz**	**2631 Hz**	**0.1918 dB**	**0.0001667**
Agm5	**660 Hz**	**903 Hz**	**1460 Hz**	**2595 Hz**	**0.1600 dB**	**0.0001667**
Agf5	**743 Hz**	**905 Hz**	**1271 Hz**	**2605 Hz**	**0.1606 dB**	**0.0001667**
Test samples						
Sound	**F1**	**F2**	**F3**	**F4**	**Amp**	**Per**
Ahd6	**796 Hz-20.5 dB**	**1239 Hz-16.5 dB**	**2206 Hz-2 dB**	**X**	**0.2603 dB**	**0.0001667**

Ahm6	733 Hz-19 dB	1213 Hz-16 dB	2627 Hz-4.5 dB	X	0.1320 dB	0.0001667
Ahf6	747 Hz-20.5 dB	1143 Hz- 17.5 dB	2684 Hz-5 dB	X	0.2278 dB	0.0001667
Ahd7	714 Hz–16.5 dB	1075 Hz-22 dB	2028H-6 dB	X	0.1638 dB	0.0001667
Ahm7	672 Hz-17 dB	1085 Hz-19 dB	2054 Hz-5.5 dB	X	0.1652 dB	0.0001667
Ahf7	565 Hz-6.5 dB	1552 Hz-5 dB	2505 Hz-11 dB	X	0.2275 dB	0.0001667
Ahd8	663 Hz-18 dB	1134 Hz-14 dB	2326 Hz-5 dB	X	0.1641 dB	0.0001667
Ahm8	663 Hz-18 dB	1219 Hz-17 dB	2234 Hz-9.5 dB	X	0.1632 dB	0.0001667
Ahf8	602 Hz-18 dB	1037 Hz-12 dB	2374 Hz-7.5 dB	X	0.1359 dB	0.0001667
Ahd9	670 Hz-14 dB	1068 Hz-13.5 dB	1750 Hz-2 dB	2641 Hz-3 dB	0.1600 dB	0.0001667
Ahm9	661 Hz-18 dB	1088 Hz-15.5 dB	2543 Hz-2 dB	X	0.0998 dB	0.0001667
Ahf9	632 Hz-17 dB	632 Hz-17 dB	1689 Hz-3 dB	2634 Hz- -1 dB	0.1312 dB	0.0001667
Ahd10	714 Hz-15.5 dB	1209 Hz-20 dB	2474 Hz-6 dB	X	0.1635 dB	0.0001667
Ahm10	691 Hz-19 dB	1167 Hz-15.5 dB	2329 Hz-9 dB	X	0.1910 dB	0.0001667
Ahf10	642 Hz-19 dB	994 Hz-16 dB	1498 Hz-5 dB	2315 Hz-1 dB	0.1626 dB	0.0001667
Afd6	466 Hz-0.5 dB	1210 Hz-17 dB	1659 Hz-9 dB	X	0.1635 dB	0.0001667
Afm6	502 Hz-1 dB	1217 Hz-17 dB	1681 Hz-9 dB	X	0.2280 dB	0.0001667
Aff6	661 Hz-2 dB	1030 Hz-20 dB	1580 Hz-18 dB	2477 Hz-6 dB	0.1970 dB	0.0001667
Afd7	777 Hz-2 dB	1082 Hz-14 dB	1604 Hz-12 dB	2528 Hz-0.5	0.0966 dB	0.0001667
Afm7	946 Hz-16.5 dB	1581 Hz-12.5 dB	2692 Hz-3 dB	X	0.1280 dB	0.0001667
Aff7	687 Hz –3.5 dB	969 Hz- 19 dB	1491 Hz-13 dB	2562 Hz-4 dB	0.1009 dB	0.0001667
Afd8	691 Hz-17 dB	971 Hz-14 dB	1387 Hz-16.5 dB	2475 Hz-8.5 dB	0.1280 dB	0.0001667
Afm8	599 Hz-6 dB	880 Hz-21 dB	1448 Hz-10 dB	2815 Hz-5 dB	0.1632 dB	0.0001667
Aff8	505 Hz-6.5 dB	1009 Hz-19.5 dB	1260 Hz-19 dB	2522 Hz-10 dB	0.1973 dB	0.0001667
Afd9	815 Hz-19.5 dB	1053 Hz-10 dB	1540 Hz-12.5 dB	2601 Hz-2 dB	0.1956 dB	0.0001667
Afm9	849 Hz-16 dB	1384 Hz-14 dB	2610 Hz-6.5 dB	X	0.1312 dB	0.0001667
Aff9	858 Hz-14.5 dB	1428 Hz-13 dB	2706 Hz—5 dB	X	0.1629 dB	0.0001667

(continued)

Table 5.10 (continued)

Referential samples						
Sound	F1	F2	F3	F4	Int	Per
Afd10	**1020 Hz-17 dB**	**1461 Hz-15 dB**	**2366 Hz-8 dB**	**X**	**0.1944 dB**	**0.0001667**
Afm10	**1041 Hz –18.5 dB**	**1599 Hz-16 dB**	**X**	**X**	**0.1312 dB**	**0.0001667**
Aff10	**752 Hz-18 dB**	**888 Hz-14.5 dB**	**1308 Hz-16 dB**	**2351 Hz-6 dB**	**0.2278 dB**	**0.0001667**
Agd6	**791 Hz-12 dB**	**983 Hz-14 dB**	**1353 Hz-9 dB**	**2274 Hz-2.5 dB**	**0.2545 dB**	**0.0001667**
Agm6	**830 Hz-15.5 dB**	**1108 Hz-14.5 dB**	**1534 Hz-15 dB**	**X**	**0.1907 dB**	**0.0001667**
Agf6	**956 Hz -19 dB**	**1330 Hz- 12.5 dB**	**2249 Hz-4 dB**	**X**	**0.1291 dB**	**0.0001667**
Agd7	**795 Hz-9 dB**	**1184 Hz-21 dB**	**1601 Hz-11 dB**	**X**	**0.2231 dB**	**0.0001667**
Agm7	**565 Hz-6.5 dB**	**1552 Hz-4 dB**	**2505 Hz-10 dB**	**X**	**0.2228 dB**	**0.0001667**
Agf7	**634 Hz-3.5 dB**	**1015 Hz-17.5 dB**	**1547 Hz-11 dB**	**2295 Hz-8 dB**	**0.2545 dB**	**0.0001667**
Agd8	**873 Hz-16.5 dB**	**1074 Hz-12.5 dB**	**1497 Hz-18 dB**	**2377 Hz-4 dB**	**0.1289 dB**	**0.0001667**
Agm8	**896 Hz-19 dB**	**1240 Hz-14.5 dB**	**1595 Hz-13.5 dB**	**2428 Hz-7 dB**	**0.1956 dB**	**0.0001667**
Agf8	**801 Hz-14.5 dB**	**1028 Hz-14 dB**	**1235 Hz-13.5 dB**	**2269 Hz-7.5 dB**	**0.1312 dB**	**0.0001667**
Agd9	**565 Hz-6.5 dB**	**1552 Hz - 4.5 dB**	**2505 Hz-10 dB**	**X**	**0.1635 dB**	**0.0001667**
Agm9	**925 Hz-18.5 dB**	**1538 Hz-9 dB**	**2373 Hz-1.5 dB**	**X**	**0.1317 dB**	**0.0001667**
Agf9	**776 Hz-16.5 dB**	**1205 Hz-11.5 dB**	**1979 Hz-0 dB**	**X**	**0.1915 dB**	**0.0001667**
Agd10	**947 Hz-18 dB**	**1318 Hz-13 dB**	**2645 Hz-0.5 dB**	**X**	**0.2205 dB**	**0.0001667**
Agm10	**918 Hz-23 dB**	**1434 Hz-11 dB**	**2684 Hz-1 dB**	**X**	**0.1439 dB**	**0.0001667**
Agf10	**581 Hz-15 dB**	**887 Hz-21.5**	**1191 Hz- 26.5 dB**	**2513 Hz-9.5 dB**	**0.1919 dB**	**0.0001667**

5.8.1 *Results and Discussion*

Many experiments represented in eight neural networks have been done in this work to show which one can bring better results (four neural networks with ordered data, while four other networks with non-ordered data). Without further ado, here are the neural networks with ordered data: neural network number 1 consists of 1 hidden layer, 2 nodes, 150 steps and the error rate of **0.0152**. The second net is of 1 hidden layer, 3 nodes, 150 steps and error rate of **0.0073**. The third net is of 1 layer, 4 nodes, 150 steps and error rate **0.0057**. The network number 4, however, contains 1 layer, 5 nodes, 150 steps and error rate of **0.0049**.

As far as the neural networks with non-ordered data are concerned, the net 5 contains 1 layer, 2 nodes, 150 steps and error rate of **0.0167**. Net 6 has 1 layer, 3 nodes, 150 steps and error rate of **0.0081**. Net 7 is of 1 layer, 4 nodes, 150 steps and error rate of **0.0066**, while net 8 has an error rate of **0.0032** with 1 layer, 5 nodes and 150 steps.

The most important thing to mention is that the neural network number 8 shows a satisfying result, i.e. has given an error rate of **0.0032**, this is because it is characterized by 1 layer, 5 nodes and 150 steps and the output dealt with are non-ordered data. This means that the random selection of data, including the use of the random simple sample represented in 20 speakers of different ages and sex, can give satisfying results in automatic processing. Simply put, the principle of automatic recognition for the syllable */ʔa/* is the same for the other syllables and therefore can be applied to Arabic and other languages in terms of syllables. Table 5.11 displays the results in number the optimal neural network has shown below, i.e. number 8 with 1 layer, 5 nodes, 150 steps and error rate of **0.00320**.

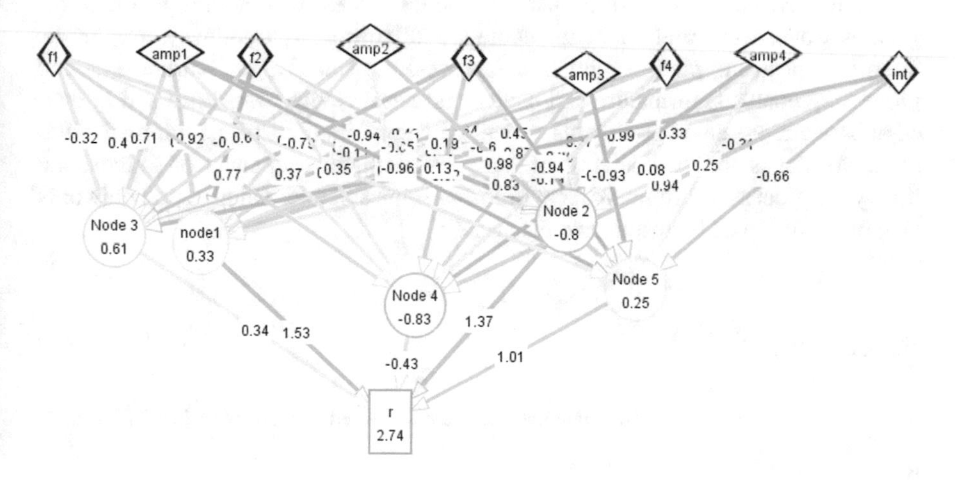

Table 5.11 Results

Number of NNW	Number of hidden layers	Number of nodes	Number of steps	Error rate
NNW with ordered data				
1	**1**	**2**	**150**	**0.0152**
2	**1**	**3**	**150**	**0.0073**
3	**1**	**4**	**150**	**0.0057**
3	**1**	**5**	**150**	**0.0049**
NNW with non-ordered data				
Number of NNW	**Number of hidden layers**	**Number of nodes**	**Number of steps**	**Error rate**
5	**1**	**2**	**150**	**0.0167**
6	**1**	**3**	**150**	**0.0081**
7	**1**	**4**	**150**	**0.0066**
8	**1**	**5**	**150**	**0.0032**

5.9 Conclusion

Although speech recognition is a thorny area, this chapter tries to give a clear insight about this field in terms of tools and techniques implemented to reach an acceptable result. The chapter, then, relates between several fields speech technology, acoustic phonetics, human intelligence and artificial intelligence.

It starts with speech recognition, fundamentals of speech recognition, artificial intelligence, neural networks, fundamentals of neural networks, neural networks as a set of processing units and neural networks as a set of relations and programming and then moves to the field of acoustic phonetics and explains some specific terms such as wave, wave motion, fundamentals and harmonics, frequency, amplitude, intensity, spectrum, pitch, sound signal processing, sound signal variance, sound processing analysis, formants, oral cavity and nasal cavity. Data base and knowledge base are presented afterwards to deal with combinations of the minimal syllable in six stages and then deletion of some unacceptable sequences of sounds, and finally the process of automatic recognition of some syllables using neural networks is applied, followed by results and discussion.

References

Arab World English Journal (2014) International Peer Reviewed Journal. 5(4). ISSN 2229-9327 AWEJ

Birjandi P, Salmani-Nodoushan MA (2005) An introduction to phonetics. Zabankadeh Publications, Tehran

Connor OJD (1973) Phonetics a simple and practical Introduction to the nature and use of sound in language. Cambridge University Press, Cambridge, Cambridge new edition. ISBN 014 013638x

De Saussure F (2002) *Cours de linguistique générale*. édition Talantikit. Bejaia

Dinedane N (1995) *L'Intelligence Artificielle Elément de Base*. Office des Publications Universitaires. ISBN 978-9961-00-001-4
Douglas BH (2007) Principles of language learning and language teaching, 5th edn. Pearson Education, White Plains. ISBN 10-0130178160
Emerit E (1977) *Cours de Phonetique Acoustique*. Sibawajh collestion dirigé par Abderrahmane Hadj–Salah Directeur de de linguistique et de phonétique de l'université d'Alger
Fant G (2004) Speech acoustics and phonetics. Kluwer Academic, Dordrecht
Fry DB (1979) The physics of speech. Cambridge textbooks in linguistics, 1st edn. Cambridge University Press, Cambridge
Gimson AC (1980) An introduction to the pronunciation of English, 3rd edn. Edward Arnold, London
Goldberg R, Riek L (2000) A practical handbook of speech. CRC, Boca Raton
Lieberman P, Blumstein S (1988) Speech physiology, speech perception, and acoustic phonetics. Cambridge Studies in speech science and communication. Cambridge University Press, Cambridge
Neffati T (1999) Génie éléctrique. Traitement de signal analogique, Technosup, les filières téchnologiques des enseignements supérieures. Ellipses Edition Marketing, Paris
Omar AM (1977) Mohadart fi Ilmi alugha Alhadith (Lectures in Modern Linguistics), Kuliyat dar alouloum, jamiaat alqahira, 1st edn. Faculty of House of Sciences, University of Cairo, Cairo
Peinado AM, Segura JC (2006) Speech recognition over digital channels robustness and standards. Wiley, West Sussex
Plomp R (2002) The intelligent ear on the nature of sound perception. Lawrence Erlbaum, Mahwah
Rabiner L, Juang BH (1993) Fundamentals of speech recognition. Prentice Hall, Englewood Cliffs
Roach P (1991) English phonetics and phonology. A practical course. Cambridge University Press, Cambridge
Roach P (1992) Introducing phonetics. Penguin, Harmondsworth
Sean AF (2011) Speech spectrum analysis signals and communication technology. ISSN 1860-4862. Springer, Berlin
Tang H, Tan KC, Yi Z (2007a) Neural networks: computational models and applications. University of Birmingham, Birmingham
Tang H, Tan KC, Yi Z (2007b) Neural networks: computational models and applications. Applications. Studies in computational intelligence. Springer, Berlin
Tebelskis J (1995) Speech recognition using neural net works. Ph.D Thesis, CMU-CS-95-142 School of Computer Science, Carnegie Mellon University, Pittsburg, pp 15213–13890
Wu Chou A (2009) Minimum classification error (MCE) approach in pattern recognition. Labs Research, Avaya Inc., Santa Clara

Arabic References

Abdeljlil A (1991) *Al Aswat Alughawiya* (Speech sounds), 1st edn. Dar Assafaa Li Al Nashr Wa Al Tawzie, Oman, Jordan
Issam N (1992) *Ilm Al Aswat Al Lughawiya Al Phonetica* (Phonetics), 1st edn. Dar Alfikr Allubnani. Fikr House, Beirut, Lebanon
Kaddour AM (1999) *Mabadie alissaniyat* (Principles of linguistics), 2nd edn. Dar alfikr Almoaasir, Beirut, Lebanon

General Conclusion

This research was about one English phonological phenomenon represented in the weak vowel known as schwa and its equivalence in Arabic, relying upon a contrastive study and a spectral analysis and trying to treat the Arabic minimal syllable automatically hinging upon the neural networks. To this account four chapters have been used. Chapter 1 was about the Arabic phonological system and dealt with impact of environment on language, Flency (Alfasaha), Arabic sounds, sounds used by Arabs, structure of Arabic syllable, types of syllables, stress, compound vowel, difference between diphthong and vowel, intonation in Arabic, functions of intonation, infrequent use of Dammah and Kasrah by Arabs, pause and patterns of nouns in Arabic. Chapter 2 dealt with the English phonological system and included language, aim of the language, language as a symbol, language as an interaction, English sounds, vowels, characteristics of vowels, types of vowels, schwa, consonants, classification of consonants, active and passive articulators, syllable, stress and intonation. Chapter 3 dealt with a contrastive study in English and Arabic and encompasses the following points: phonemic analysis, definition, taxonomic phonemics, sounds, English corpus, Arabic corpus, inventory of words containing schwa, inventory of words containing ḫ sound in final position, inventory of schwa and h sound in final position, words in contrast, segmentation to show contrast, spectrum and spectral analysis. Chapter 4 encompasses the following: speech chain, speech technology, discrimination between speech sounds, speech recognition, fundamentals of speech recognition, intelligence, artificial intelligence, neural networks, fundamentals of neural networks, neural networks as a set of processing units, neural networks as a set of relations, programmation, acoustic phonetics and signal processing, database, knowledge base, deletion by opposition, stages of combination, combination of the minimal syllable, syllable recognition using neural networks and results and discussion.

M. Dib, *Automatic Speech Recognition of Arabic Phonemes with Neural Networks*, SpringerBriefs in Applied Sciences and Technology,
https://doi.org/10.1007/978-3-319-97710-2

The obtained results were as follows:

- Schwa located in final position is the equivalence of pause in Arabic which is represented in an open vowel and fricative consonant (h).
- Theory. of transfer one of the theories of contrastive linguistics is efficient for teaching English to Arab speakers.
- Generative approach for Arabic sounds provides a very large corpus.
- The use of the random simple sample is represented in 20 speakers of different ages, and sex gives good results.
- The optimal neural network is that of a non-ordered data base, which consists of 1 layer, 5 nodes and 150 steps.
- The random selection of data can give satisfying results in automatic processing.
- The principle of automatic recognition for the syllable ***/ʔa/*** is the same for the other syllables.

As recommendations, we suggest researches into other phonological phenomena in English contrasting them with some Arabic phonological phenomena to ease the task for teaching English to Arab speakers; this is because one cannot understand the value of a particular thing without doing contrast, and researches in speech recognition are relying upon the neural networks of five nodes with the random selection of data because this can help tremendously in automatic speech recognition.

Moreover, although the principle of automatic recognition has been applied to one phonological unit concerning the minimal syllable ***/ʔa/***, it is necessary to enlarge the corpus, i.e. working on all the Arabic phonemes and syllables and trying to treat all syllables in Arabic automatically so that computer specialists will be able to join them and convert speech into manuscript.

Finally despite the great effort devoted to this research, it is considered an attempt to strike two birds with only one stone, to study one of the English phonological phenomena called schwa and to recognize one type of Arabic syllable automatically called minimal syllable. Much work still needs to be done to explore other English phonological units, which seem ambiguous with regard to the Algerian learners and trying to contrast them with English so as to ease the task for English foreign learners and to do researches in speech recognition so as to boost Arabic in the field of artificial intelligence.

Appendices

Phonetic Transcription of the English Text

vıdıəu ðə mæn wıð ðə baıɒnʃk aı

bı bı sıınsaıd aut lɒndən

mɑ :ʧ tu: Θauzənd ænd naın

s^bʤəkt hau saınsfıkʃn ız bıkɒmıŋə rıælıtıwıð ə rımɑ:kbl medıkl breıkΘru: fə blaınd

baıənık aı

lɒndən aı hɒspıtl ız æt ðə fɔ:frɒnt ɒv ə junık traıl ðæt hæz ðə pətenʃl tu rıstɔ: blaınd pi:plz saıt. bı bı sıınsaıd aut lɒndən wɒz gıvən ıksklusıv ækses tu grandbreıkıŋ wɜ:k teıkıŋ pleıs ætmu:fıldz aı hɒspıtl.

məustɒv ^s teık auə saıt fə grɑ:ntıd b^t tu mılıən pi:pl ın ðə ju keı hæv sɒm sɔ:t ɒv prɒblm wıð ðeə saıt.

fə twentı faıv Θauznd ın lɒndən blaındns ænd kəndıʃn retınıs ız ə deılı rıælıtı.

æt mu:fıldz aı hɒspıtl sɜ:ʤənz paıənıərıŋə junık ıvenʃn ðæt ıneıblz sɜ:ʤənz tu ft peıʃnts hu hv lɒst ðeə saıt wıð ə baıɒnʃk aı.

təutl dɑ:kıns

rɒnız wɒn ɒv ðə kæptlz twentı faıv Θauznd blaınd rezdnts ænd bıkəz ɒv hıredıtərı kəndıʃn hi: hæz bi:n lıvıŋın təutl dɑ:knıs fə læst Θɜ:tı jıəz

hi: ız nau wɒn ɒv ʤ^st Θri: pi:pl ın ðə k^ntrı tu hæv bi:n fıtəd wıð ə revəlu:ʃnərı baıɒnʃk aı wıʧız hævıŋə drəmætık ımpækt ɒn hız laıf.

Θænks tʊ ðıs ımplɑ:nt hi: ız nau eıbl tu si: dıfrənt ʃeıdz ɒv laıt.hi: kæn nau wɔ:k ə lɒŋə waıt laın peıntıd ɒn ðə graund ænd ıvən sɔ:t aut hız sɒks ıntu: waıt blæk ænd greı paılz.

Laıtıŋ ^p rɒnz wɜ:ldɒv dɑ:knəs

baıənık aı wɜ:ks baı kæpʧərıŋ laıt ɒntu vıdıəu kæmrəın peıʃnts glæsi:z wıʧ sendz waıələs sıgnəl tu ðəımplɑ:nt wıʧ stımjuleıts ðəɒptık nɜ:v

M. Dib, *Automatic Speech Recognition of Arabic Phonemes with Neural Networks*, SpringerBriefs in Applied Sciences and Technology,
https://doi.org/10.1007/978-3-319-97710-2

ðɪs medɪkl breɪkru: hæz ðə ptenʃl tu nɒt ʤ^st rædɪklɪ trænsfɔ:m ðə laɪvz ɒv blaɪnd b^t ɔ:lsəu tu ɪmpruv nɔ:ml saɪt pauəz ɒv fjuʧəʤə nəreɪʃnz

ðɪsɪz nɒt əuvə naɪt mərɪkl kjuə tu blaɪndnəs æt ðə məumənt ɪts əunlɪ beɪŋ traɪəld ɒn ə verɪ spəsɪfɪk grup

wɒtɪz səu ɪksaɪtɪŋəbaut ðə baɪənɪk aɪɪz ɪts ptenʃl fɔ: ðə fjuʧə

grəgɔ: kəsendeɪ frɒm səknd saɪt sez ɪn fɪvtɪ ji:z taɪm aɪ həup ðæt pi:pl wɪl bi: eɪbl tu ri:d wɪð ðɪs sɪstəm ænd ɪtsnɒt ^nΘɪnkəbl ðæt ɪn ðə dɪstənt fjuʧə pi:pl wɪl hæv retɪnəɪmplɑ:nt ðæt kæn prəvaɪd ðəm wɪð betə vɪzn ðæn nɔ:ml sɪjɪŋ pi:pl.

bɪ bɪ sɪɪnsaɪd aut ɑ:skt peɪʃnt rɒn əbaut hɪz ɪkspɪrjəns wɪð ðə baɪənɪk aɪ

prɪzentə mæju: raɪt ænd rɒn

haʊɑ: ju: getɪŋəlɒŋ

rɒn sləʊlɪ b^t ʃuəlɪ ðeɪ sæd let ðeə bɪ laɪt ænd ðeə wɒz laɪt

fəΘɛ:ti:n jɪəz av si:n æbsəlutlɪ nɒΘɪŋ æt ɔ:l ts ɔ:l bi:n blæk bt naʊ laɪt ɪz kɒmɪŋΘru:

ɪt ɪz trʊlɪəmeɪzŋ ðeɪ r wɒndəfl pi:pl i:z saɪntɪsts ɪts ɪksaɪtɪŋ æftə ju:v si:n nɒΘɪŋ fəΘɛ:ti:n jɪəz b^t dɑ:knəss^dənlɪ tʊ bɪ eɪbl tʊ si: laɪt əgen ɪz trʊlɪ wɒndəflɪts laɪk ðe fjuʧə kɒmɪŋ tʊ ^s naʊɪn ðəprezənt ɪzntɪtrɒnɪtɪz maɪ wɒn æmbɪʃn æt ðə məʊmənt ɪz tʊ si: ðə mu:n tʊ gəʊ aʊt ɒn ə naɪs klɪə i:vnɪŋ ænd tʊ bɪ eɪbl t pɪk ^p ðə mu:n weðə aɪl bɪ eɪbl tʊ dʊɪt ɔ: nɒt aɪ dəʊnt nəu b^t æm rɪlaɪŋɒn ði:z saɪntɪsts

haʊ ðɪd ju: lu:z jɔ: saɪt rɒn

rɒn ɪts ə fæməlɪΘɪŋɪts wɒn ə ði:z hərɪdɪtrɪ kəmpleɪns kɔ:ld retɪnəs pɪgməntəʊsə nɔ:məlɪ nəʊn æz t^nəl vɪʒn ænd beɪsɪklɪ jɔ: pərɪfərl vɪʒn stɑ:tstʊ dɪsəpɪə ^ntɪl jɔ: left wɪð sentrəl vɪzn wɪʧ mi:nz ju: kən rɪkəgnaɪz s^mw^n fɪvtɪ jɑ:dz daun ðə rəud ænd weɪv tʊ ðəm ænd wɔ:k ɪntʊə læmpəust wɪʧɪz əʊnlɪ sɪks ɪnʧəz æt jɔ: saɪd ænd ðen ɪventju:lɪ maɪ sentrəl vɪzn went ænd aɪ wəz rəʤɪstəd blaɪnd ɪn naɪnti:n sevəntɪ naɪn. aɪ gɒt ə gaɪd dɒg ɪn naɪnti:n eɪtɪ ænd aɪv nevə lu:kt bæk frəm ðæt.

ɪn tɜ:mzɒv jɔ: saɪt naʊ ju: hæv nəʊ saɪt æt ɔ:l

rɒn nɒn wɒt səʊ evə evrɪΘɪŋɪn blæk ðə ri:zn aɪΘɪnk ðæt ɪt tʊk mi: ə lɒŋ waɪl tʊ meɪk ^p maɪ maɪnd weðə aɪ wɒntəd tʊ gəʊ fə ðɪs eksperɪmənt bɪkɔ:z ɪt ment Θri: ə fɔ: aʊəɒpəreɪʃn ænd ju: nəʊɒbvju:slɪ ju: weə r ɪn hɒspɪtl ænd ðə saɪntɪsts dɪdnt nəʊɪgzæktlɪ wɒt ðə rɪz^lts weə gəʊɪŋ tʊ bi: ænd ju: dɪdnt …

haʊ dɪd ju: fɜ:st hɪə r əbaʊt wɒt wəz gəʊɪŋɒn æt mu:fi:ldz

rɒn beɪsɪklɪΘru: maɪ waɪf ʃi: ju:zd tʊ wɜ:k fə ðə gaɪd dɒg əsəʊsɪeɪʃn ænd ʃi: ki:ps ^p tʊ deɪt wɪð retɪnəs pɪgmentəʊsəəsəʊsɪeɪʃn ænd wi: get ə mægəzɪn evrɪ kwɒtə.

ðeɪ menʃnd ðæt ðeə wəz gəʊɪŋtʊ bi: ə semɪnɑ: held fə ðɪs ədvɑ:nst teknɪk

baɪ sekənd saɪt. səʊʃi: tʊk ðə dɪteɪlz ænd ɪventju:lɪʃi: pɜ:sju:weɪdəd mi: ðæt aɪʃu:d æt li:st gəʊəlɒŋg ænd lɪsən tʊ wɒt wəz beɪŋ sed… bɪfɔ: aɪ nju: weə aɪ wəz ʃi:d put maɪ neɪm ɒn ðə lɪst.

wɒns jɔ: waɪf put jɔ: neɪm fɜ:wəd wɒt wəz ðə prəʊsəs.

rɒn ju: weə r ɪntvju:d ænd ju: hæd tʊ kəmplaɪ wɪð faɪv kraɪti:rɪə… aɪΘɪnk ju: hæd tʊ bi: kəmpli:tlɪ blaɪnd ju: hæd tʊ lɪv wɪðɪn tu: haʊəz ɒv mu:fi:ldz.

ɒVɪəslɪ ju: hæd tʊ bi: ebl tʊ kənveɪ tʊ ðə saɪntɪsts wɒt ju: kʊd si: ænd jɔ: gæŋglɪənz hæd tʊ bi: ɪn ɔ:də

dɪd ju: hæv kənsɜ:nz əbaʊt æn ɒpreɪʃn

rɒn nəʊ ɪf æm æbsɒlu:tlɪ hɒnɪst wɪð ju: ɪts ɔ:lweɪz bi:n maɪ aɪdɪə ðæt daɪjɪŋ kæn bi: peɪnfl ænd ðə wɒn weɪ ju: kæn ɪlɪmneɪt penfl deΘɪz baɪ gəʊɪŋ ænd ^ndəɒpəreɪʃn bɪkɔ:z ju: nəʊ nɒΘɪŋəbaʊt ɪt

wɒt dɪd ðə ɒpəreɪʃn ɪnvɒlv rɒn

rɒnɪts ðə raɪt aɪ ðeɪɒpəreɪt ɒn. ðeɪəʊpn ðɪ aɪ ænd ðeɪɪmplɒnt ə smɔ:l ə reɪ ænd ðeɪ tæk ɪt tʊ ðə bæk ɒv ðə retɪnə ænd ɪt kənteɪnz sɪkstɪ taɪnɪəlektrəʊdz

i:ʧɒv ðəʊz əlektrəʊdz ɪz kənektəd tʊə waɪə ænd ðæt waɪə r ɪz brɔ:t aʊt frəm ðə saɪd ɒv ðə aɪ bɪləʊ ðəʧekbəʊn weə ju: kɑ:nt si: ɪt

let ðeə bi: laɪt aɪɒpəreɪʃn

ə lɪtl reɪdɪəʊ resi:və fə wɒnt ɒv ə betə wɜ:d ɪz pleɪst ðeə ænd ə pi:s ɒv dɒneɪtəd skli:rə ðə waɪt ɒv ən aɪɪz ju:zd æz ə sɔ:t ɒv beltəd əkrɒs ðə aɪ tʊ həʊld ɪt ɪn pleɪs.

ðen ju: k^m əkrɒs ðə glæsəz wɪʧ kənteɪn ə lɪtl kæmrəɪn ðə nəʊz pi:s ænd ɪt ɔ:lsəʊ hæz ə reɪdɪəʊ lɪŋk

ðeɪ kɔ:l ɪt ən ɑ: ef lɪŋk wɪʧɪz ətæʧt tʊ ðəʧi:k wɪʧɪz ɒn ðə glæsəz b^t prest əgenst ðəʧi:k ænd ə keɪbl r^nz frəm ðə kæmrə tʊə smɔ:l kəmpju:tə wɪʧ ju: kən weə (r) ɒn ə belt nəʊ bɪgə ðæn ə pækɪt ɒv sɪgrəts

ðen ði ɪnfəmeɪʃn ðæt ðə kəmpju:tə resi:vz ɪz fed ^p tʊ ðəɪnd^kn kɔɪl æz ðeɪ kɔ:l ɪt ðə saɪz ɒv fɪvtɪ pens pi:s ɒn ðəʧi:k bəʊn

ðə reɪdɪəʊ sɪgnəlz trənsmi:təd tʊ ðə lɪŋk ɒn ðɪ autsaɪd ɒv ðɪ aɪ ðen ɪts baɪ ðə keɪbl tʊ ðə sɪkstɪəlektrəʊdz æt ðə bæk ɒv ðɪ aɪ wɪʧ wen ðeɪ(r) əʤɪteɪtəd ɔ: lɪt ^p meɪk ðə retɪnə rɪspɔ:nd səʊ ju: kən ækʧju:lɪ pk pɪk ^p laɪt

enɪ rɪgri:ts

rɒn nəʊɪts ə greɪt prɪvəlɪʤ ænd ən hɒnə tʊ bi: eɪbl tʊ teɪk pɑ:t ɪn ən ɪksperɪmnt s^ʧ æz ðɪs həʊpɪŋ ðæt ðɪ autkəm ɪz gəʊɪŋ tʊ bi: eɪbl tʊ brɪŋ saɪt tʊ pi:pl laɪk maɪ self ðæt ɑ: kəmpli:tlɪ blaind

wɒt dʊ ju: si: wɒt kən ju: si:

rɒn ðə wɒn ədvɑ:ntɪʤɪt hæz æt ðə məʊmənt ɪz mɔ: ɒn maɪ waɪfs saɪd bɪkɔ:z aɪ kən naʊ sɔ:t ðə wɒʃɪŋ aʊt ɪt gɪvz mi: greɪdz ɒv braɪt laɪt tʊ blæk ænd enɪΘɪŋɪn bɪtwi:n

aɪ kən ækʧju:lɪ sɔ:t aʊt waɪt sɒks greɪ sɒks ænd blæk sɒks b^t æz fɑ: z wɒʃɪŋɪz kənsɜ:nd ɪts ʤ^st ə kweʃn ɒv Θɪŋz ɑ: aɪðə waɪt ɔ: ðeɪ (r) k^ləd ænd ðæt sju:ts maɪ waɪf daʊn tʊ ðə graʊnd

kæn ju: si: ʃeɪps

rɒn ðeə (r) ɪz nəʊʃeɪps æz fɑ: z æm kənsɜ:nd ju: kɑ:nt si: prɪnt ɔ: enɪΘɪŋ laɪk ðæt

vɪʒn æz ju: nəʊ ɪt ɪznt ðeə aɪ kən pɪk ^p ə wɪndəʊ aɪ kən pɪk ^p pɒsɪblɪ ə dɔ: freɪm b^t æz fɑ:z fə ju:zfl vɪzn tʊɪneɪbl mi: tʊ mu:v əraʊnd k^mftəblɪ aɪ l stɪk tʊ ðə gaɪd dɒg

b^t ðɪs ɪz ɜ:lɪ deɪz ɪts əʊnlɪ sɪks m^ns

wɪl ðɪs ɪmpru:v ɪn ðə fju:ʧə

rɒn aɪ sɪnsɪəlɪ həʊp səʊ aɪΘɪŋk ɪts rɪəlɪ wɜ:kɪŋ wɪð saɪntɪsts ænd ɔ:lsəʊ edjukeɪtɪŋ jɔ: breɪn tʊ ^ndəstənd wɒt jɔ: si:ɪŋ

æt ðə məʊmənt aɪ ni:d s^mw^n tʊ tel mi: wɒt æm lʊkɪŋ æt b^t æm həʊpɪŋ ðæt maɪ breɪn wɪl bɪgɪn tʊ pʊt ðə pɪkʧəz tʊgeðə tʊɪneɪbl mi: tʊ ^ndəstænd wɒt æm lƱkɪŋ æt

Phonetic Transcription of the Arabic Text

Sarata:n alkıla: ʔasa:li:b ʕıla:ʒıjah hadi:θah

ʕıla:ʒa:t faʕalah tamnahu lmardıa: amalan kabi:ran fi: haja:tıʔatwalılmuxtasıar. juʕtabaru Sarata:n alkıla:mına lmuʃkıla:tı asıhıjah alqa:tılahfahıuwa lmusabıbu raʔi:si: aΘa:nı lılwafa:tı fi: alʕa:lam wa tuʃi:ru lʔıhsa:ʔıja:t alxa:sah bısarata:n alʒıha:zı lbawlı wa tana:sulıʔıla: ʔana hıuna:ka zıja:dah sanawıjah fi: mardıa: sarata:nı alkılah fi: alwatanı alʕarabıj muwa:kabatan lızıja:datı lha:sılah fi: lʕa:lam wa laqad wuʒıda ʔana ha:la:tı sarata:n lkılah qad za:dat bınısbatıʔıΘna:nı wa xamsu:na fi: lmıʔah xıla:la alfatra bajna ʕa:maj ʔalf wa tısʕu mıʔa wa Θala:Θa wa Θama:nu:n wa ʔalfajnı wa ʔıΘnajn ʔaj mın sabʕa fa:sılah wa:hıdʔıla: ʕaʃarah fa:sılah Θama:nıjah ha:lah lıkulı mıʔata ʔalf ʃaxsı

wartafaʕat muʕadala:t alwafaja:t ʔajdıan wa bıʃaklın xa: sın bajna ʔu:la:ʔıka lmu sıa:bi :na bılʔawra:mı lʔakbaru haʒman mın sabʕatu santımıtra:tın fartafaʕat mın wa:hıd fa sılah Θala :Θah ʔıla: Θala:Θah fa sılah ʔıΘna:n lıkulı mıʔata ʔalf ʃaxsıkama: tuʃi:ru kaΘi:run mın alʔıh sıa:ʔıja:t alʕarabıjah ʔıla: zıja:datı nısbatı mardıa: lkıla: fi: lwaṭan lʕarabı famaΘalan tubajınu ʔıhsıa:ʔa:t asıʒılı lwaṭanı asaʕudi:lılʔawra:m ʔana muʕadal lʔı sıa:bah bısaraṭa:nı lkıla: fi: lmamlakatı alʕarabıjah asaʕu:dıjah wa sıal ʔıla: mıʔataj ha:lah sanawıjan wa akΘaruha: mın maka lmukaramah wa rija:dı wa lmanṭıqatıʃarqıjah wa lılwʊqu:fıʕala: ʔaḥamı lmʊstaʒada:tı lʕılmıjah wadawa:ʔıjah lısaraṭa:nı lkıla: ʔaqa:mat alʒamʕıyah saʕu:dıjah lıʒıra:hatı almasa:lıkı lbawlıjah bıtaʕa:wʊnı ma a alʒamʕıyah saʕu:dıjah lılʔawra:mı mʊʔtamaran fi: madi:natıʃaramıʃajxı fi: lfatratı mın ʕaʃara ʔıla: ıΘna: ʕaʃara fıbra:jar alha:li: wa ḍımna hadi:Θıḥı lışhatık awdaha: raʔi:sʊ lmʊʔtamar alʔʊsta:ð adʊktu:r ʔaʃraf abu: samrah ʔıstıʃa:ri: ʒıra:hat awra:mı almasa:lık albawlıjah madi:nat almalık ʕabdʊ lʔazi:z aṭıbıjah fi:j ʒadah

ʔına lḥadafa lʔasa :asi : lımuʔtamar ḥuwa rafʕu mustawa : lwaʕjı lada: lʔaṭıbaʔı wa bıtabaʕıja lada: lmarḍa: walıḥtıma:mı bıtarsi:xı qa:ʕıdat alwıqa :jatu xajrun mınalʕıla: ʒı wa lʔıʃa :ratıʔıla: ʔaḥamıjatı

waxuṭu:ratı maraḍı saraṭa:nı lkıla: wa xuṣuṣan baʔda zıja:datı nısbatı lʔısa :batı bılmaraḍı fi: lʔa:lamıʔaʒmaʕ wa ʔakada aduktu:r ʔabu: samra ʕala: lʔaʃxa:ṣı laði :na ladajḥım ta:ri:xa marda: ʕa:ʔıli: lısaraṭa:nı lkıla: ʔan juxbıru: ʔaṭıba:ʔaḥum bıðalıka wa jaqu:mu: bʕamalı lıxtıba:ra:tı la:zımah bıṣu:ratın mutakarırah falqıja:mu bıdawrın ʔıʒa :bıjın nahwa lmuha:faḍah ʕala: ṣıhah jazi :du mın furaṣıktıʃa:fı almaraḍı mubakıran ʕıla:ʒa:t faʕa:lah

juḍi:fu aduktu:r ʔabu: samra ʔana ḥuna:ka xıja:ra:tın kaΘi:rah muta:hah lıʕıla :ʒı maraḍı sarata:nı xala:ja: alkıla: wa jumkınu ʔıʕta:uʔakΘara mın nawʕın mına alʕıla:ʒa:t fi: alwaqtı nafsıḥʔıʕtıma:dan ʕala: lmarhalatı almaraḍıjah lısarata:n

faḥuna:ka ʒıra:hatu lıstıʔsa:lı alʒadri: lılkıljah ıstıʔsa:lu alkıljah bılmınḍa:r ʔıza :latu aΘa :nawja:tı almuntaʃırah alʕıla:ʒ wa ḥaða alʔaxi:r alʔʃʕa:i: alʕıla:ʒ alhajawi: wa alʕıla:ʒu bılʔadwıjatı almustaḥdafah wa lmuwaʒahah juʕadu ba :rıqata ʔamal lḥa:ʔu :la:ʔı lmarḍa: hajtu uΘbıtat faʕa :lıatuḥ fi: ʕıla:ʒı ha :ða anawʕu mına saratana :tı mıΘla ʕaqa :r wa lʔadwıjah lʔuxrah lmuma :Θılah mına nawʕı nafsıḥ xıja:ra:t mutaʕadıdah alʒɪra:hah fi : lxat̞ɪ lʔawalɪ fi: ʕɪla:ʒɪ sarat̞a:nɪ xala:ja lkɪla : maʕa kaΘi:r mɪna lmard̞a: wa lah̞a: ɪhtɪma:lɪatu ʃɪfa:ʔɪʔɪla: ʔana sarat̞an mutawasɪt̞ʔaw ʕa:li: lxut̞u:rah ɤa :lɪban ma: jaʕu:du baʕda lʒɪra:hah fi: xamsatun wa ΘalaΘi:na ʔɪla: xamsatun wa sɪtu:na fi: lmaʔah mɪna lha:la:t wa h̞aðɪh̞ɪ lhaqi :qah muhadadah bdra :sat taljah lati : amlat alfun wa stu maah waw a :hdun wa sabu :na mari :dan ladajhum ɪklɪni:kɪjan sarat̞a:n xala:ja : lkɪla: mɪn nawʕɪ lxalɪjah lwa :dha mawdi : wa fi : ʒa:nɪbɪn wa:hɪd mɪna lʒɪsmɪ xadaʕu: lɪʒɪra:hatɪ stɪʔsa:lɪn ʒadrɪjɪn wa qad hadaΘa ɪntɪʃa:run lɪlmarad̞ fi: ʔalfɪn wa sɪtu maʔa wa Θala:Θa wa sɪtu:n lɪkɪla: lsm fi: ʔarbaʕu maʔa wa tɪsʕatun wa sabʕu:na mari:d̞an fi: xɪla:lɪ wa:hɪd fa :s̞lah Θala:Θah sanah fi: lmutawas̞ɪt wa ka:na mutawas̞ɪtu fatratɪ lbaqa:ʔɪ bɪdu:nɪ ntɪʃa:r lɪlwaram sɪtatun wa Θama:nu:na fi: lmaʔa ʕɪnda sanatɪn wa:hɪdah wa sabʕatun wa sabʕu:na fi: lmaʔa ʕɪnda Θala:Θatɪ sanawa:t wa sabʕatun wa sɪtu:na fi: lmaʔa ʕɪnda ʕaʃrɪ sanawa:t amalun ʒadi:dun jaqu :lu markaz nʒi:mi:i:n at̞ɪbi: bɪʒa:mɪʕat ra:dju:d fi: h̞u:landa: ʔɪna

lbru:fɪs̞u:r bi:tar mu:ldarz mɪna lmaʒmu:ʕa:t lmuntaqa:t mɪna lmard̞a: ladajh̞a: ɪhtɪma:lun daʕi:fun lɪtat̞awurɪ lmarad̞ baʕda ɪstɪʔsa:lɪ lkɪla: wa ʔɪna lʕɪla:ʒa lmusa:ʕɪd baʕda lʒɪra:hah̞ bɪwa:sɪtatɪ alʕɪla:ʒɪ lʔɪʃʕa:ʕi: wa lʕɪla:ʒɪ lh̞urmu:ni: wa lʕɪla:ʒɪ lkɪmja:ʔi: almaʕh̞u:dɪ wa si:tu:ki:na:t að̞h̞ara kafa:aʔa daʕi:fah fi: ʕɪla:ʒɪ sarat̞a:n xala:ja: lkɪla: wa ʕala: raɤmɪ mɪn ða:lɪk fah̞una:ka ʔmalun ʒadi:dun wa jud̞i:fu ʔɪna alʔadwɪjah almuwaʒah̞a mɪΘla muΘabata:t ti:ru:zi:n ki:na:z ʕan t̞ari:qɪ alfamɪʒajɪdatu tahamulɪ wa faʕa:lah fi: ʕɪla:ʒɪ sarat̞a:n xala:ja: lkɪla: wa tuna:sɪbu alʕɪla:ʒu tawi:lɪ lʔamadh̞a:ðɪh̞ɪ lʔadwɪjah jumkɪnuh̞a: bɪta:li: ʔan tuwafɪra xɪja:ra:t lɪlʕla:ʒɪ fi: lmustaqbal baʕda lʒɪra:hah̞ wa lmarhalatu ta:lɪjah̞ mɪna taʒa:rɪbɪ sari:rɪjah̞ maʕa h̞a:ðɪh̞ɪ lʔadwɪjah̞ wa taʕni: ʔaʃu:r h̞ɪja qajda tanfi:ðɪ ha:lɪjan bɪma: fi: ð:alɪka taʒrɪbat ʔaʃu:r ʔaw suni:ti:ni:b ɪstɪxda:m su:ra:fi:ni:b kaʔɪla:ʒɪn musa:ʕɪd fi: ha:la:t sarata:n lkɪla: ða:t nati:ʒa ɤajrɪ lmuwa:tɪja wa taʕni: muqa:ranat su:ra:fi:ni:b maʕa lʕaqa:r wa taʒrɪbat s̞u:rs̞ fi: lmard̞a: laði:na juʕa:nu:na mɪn sarat̞a:nɪ lxala:ja: lkɪlawɪja alwah̞mi:alʔawali: baʕda ɪstɪʔs̞a:lɪh̞ɪ

muba:ʃaratan wa ʔaʃa:ra duktu:r mu:ldarz ʔɪla: ʔanah̞u ma:za:lat h̞una:ka ha:ʒah̞ʔɪklni:kɪjah̞ wa:d̞ɪha lɪʕɪla:ʒɪ lmusa:ʕɪd alhalu lʔamΘal jaqu:l bru:fɪsu:r ju:rʒi:n ʒi:ʃu:nd mɪn markaz ri:ktas dar ʔi:za:r t̞ɪbi: bɪʒa:mɪʕat mju:ni:x lfanɪjah̞ bɪʔalma:nɪja h̞una:ka ʕadadan mɪna lʔɪstra:ti:ʒɪa:t almuta:hah̞ ha:lɪjan wa tahta tatwi:r mɪn ʔaʒlɪ tahsi:nɪ nata:ʔɪʒ lʕɪla:ʒ wa jaʃmulu ða:lɪka tahdi:du lʒurʕatɪ lʔamΘal ʕɪla:ʒu lʔaʕra:d̞ɪ lʒa:nɪbɪjah̞ wa lmazʒɪ wa taʕa:qub alʔamΘal lɪlʕɪla:ʒa:t wa jud̞i:fu ʔana ɪstɪxda:m mazi:ʒ mɪna lʔadwɪja lmuwaʒah̞a jumkɪnu ʔan juhasɪn anaʃa:t̞ alʔɪklɪni:ki: bɪstɪh̞da:fɪ masa:r lʔɪʃa:ra:t ða :tɪh̞a: ʕala: mustawaja :t mutaʕadɪdah̞ʔama mazʒu lʕɪla:ʒɪ lmuwaʒah̞ bɪlʕɪla:ʒɪ lmana:ʕɪ

wa lkɪmja:ʔi: maʕa lʕɪla:ʒa:t lmuwaʒah̞a lʔuxra: fah̞uwa qajdu lbahΘ wa ma:ða: ʔanɪ lʕɪla:ʒɪ lmytaʔa:qɪb h̞al h̞uwa lʔafd̞al ʒɪʃu :nd ʔɪna dalaʔɪla lha :lɪja taftarɪd̞u ʔana lɪstɪxda:m almutaʕa :qɪb lmuΘabɪta:t muʕa :mɪl numuwa lʒɪda:r jaqu :l duktu:r jumkɪnu ʔan jaku:na mufi:dan mɪn du:nɪʔan juʕi:qa lʔɪstɪʒa :ba ta:lɪjah̞ vi: ʔi: dʒi: ef ada:xɪli: lɪlʔawʕɪja damawɪjah̞ wa ʕalajh̞ faʔɪna lfatrah̞ em ti: əʊ a:lɪmustah̞dafɪ ra:ba:mi:si:n lɪΘadɪja:t ada:xɪli: lɪlʔawʕɪjah̞ damawɪjah̞ qabla vi: ʔi: dʒi: ef alqus̞wa: lɪtahqi:q alfa:ʔɪda lʔɪklɪni :kjah̞ jumkɪnu an tatɪma bɪstxda:m ʔɪΘnajnɪ mɪn muΘabɪta:t ʕɪla:ʒa:t muwaʒah̞a ʔama ʕan dawrɪ lʕɪla:ʒa:t muwaʒah̞a em ti: əʊ a:

tahwi:l tari:qatu lʕamal ıla: muΘabıta:t fi: ʕıla:ʒı sarat̞:nı xala:ja: lkıla: famaʕa bıda:jatı alqarn alwa:hıdı walʕıʃri:n taɤajara ʔuslu:bu ʕıla:ʒı sarat̞:nı xala:ja: lkıla:lmuntaʃıra bıs̞u :ratın ʒadrıja nati :ʒata lmaʕrıfata Θa:qıba alʒadi:da lʕılmı lbıju:lu:ʒija: alʒuzʔıja bınısbatı lılʔawra:m wa ʔajdan lıxtıja:ra:tı lʕıla:ʒı alʒadi:da juqa:wımu lʕıla:ʒu lkımja:ʔi: bıs̞uratın kabi :rah̞ wa lıʔana sarata :nu xala:ja: lkıla: lmuntaʃır tama stıxda:mu astu:ki:na:t lıʕıla:ʒı alha:la:tı lmutaqadımah̞ wa ʕala: raɤmıı mın ða:lık faʔına h̞a:ðıh̞ı lʕıla:ʒa:t muʔaΘıra fi: ʕadadın mahdu:dın mına lmard̞a: faqat wa ɤajru muna:sıbah̞ lıɤa:lıbıjatı lmarda: bısabab sımjah̞

awalu nawʕın mına lıʕıla:ʒa:tı lmuwaʒah̞a lati: tama tas̞mi:muh̞a: ̞lıtaΘbi:tı mustaqbıla:t ati:ru:zi:n ki:na:z lati: juʕtaqadu ʔanah̞a: muh̞ımah̞ lınumuwı lwaramı wa lʔawʕıja damawıjah̞ lmuɤadıja lah̞u ka:na huwa su:ra:fi:ni:b laði: tama ʔıha :zatah̞u lıʕıla:ʒı sarat̞a:n xala:ja: lkıla: almutaqadımı fi: ʔalfajnı wa xamsah̞ bılwıla:ja:tı lmutahıdah̞ alʔamıri:kıjah̞ wa fi: alʔıtıhadı lʔuru:bi: ʕa:m ʔalfajnı wa sıtah̞ wa tama ʔıha:zatu muΘabıt̞ ati:ru:zi:n ki:na:z snıtıtıni:b fi: ʔalfajnı wa sıtah̞ wa muΘabıt̞ mustah̞daf ra:ba:mi:si:n lıΘadıja:t ti:msi:ru:li:ms muΘabıt̞ muʕa:mıl numuwıʒıda:rı lʔawʕıjah̞ bi :fa :si :zu :na :b ʔıla:ʒa:nıbı alʔıntrıfi:ru:n lıʕıla:ʒı sarat̞a:n xala:ja: lkıla: almutaqadımı fi: ʔalfajnı wa sabʕah̞ wa qad saʔalna: lbru:fi:su:r zja:ki:rka:li: mın kulıjatı t̞ıbıʒa :mıʕatı du:ka:z ʔahlu:l bıʔızmi:r fi: turkıja: kajfa jumkınu lʔıxtıja:r maʕa wuʒu:dı h̞a:ðıh̞ı lʔıxtıja:ra:tı lkaΘi:rah̞ faʔaʒa:ba ʔına kulu mari:d̞ın jaxtalıfu ʕanı lʔa:xar wa ʕala: ða:lıka faʔına ʕıla:ʒan wa:hıdan lan jufi:da ʒami:ʕu lmard̞a: wa ʔaʕt̞a: mıΘa:lan ʕala: ða:lıka almard̞a: lmusıni:n wa bına:ʔan ʕala: tılka lhaqi:qah̞ faʔına lʒamʕıjah̞ dawlıjah̞ lısarat̞a:nı lmusıni:n tu:s̞i: bılʔa:ti: ʕında lʔaxðı fi: lʔıʕtıba:r ʔansabu lʔadwıjah̞ lılʔıstıxda:mı maʕa mari:d̞ın bıʕajnıh̞ı jaʒıbu lʔaxðu fi: lʔıʕtıba:r baja:na:t asımjatı lxa:s̞ah̞ lıkulıʕıla:ʒın muwahıd ʕala: hıdah̞ wa ʔajd̞an wuʒu:dı marad̞ın mus̞a:hıbın muhadad wa ʕala : ða:lıka fakulu lbaja:na:t mına dıa :sa :tı lʕaʃwa:ʔıjah̞ wa tahlıla:tı lfarʕıjah̞ wa bara :mıʒa lwus̞u:lı lılmard̞a: almuwasıʕah̞ wa dıra:sa:tı bıʔaΘarın raʒʕi: wa dıra:sa:tı lha:lah̞ wa lxıbra lʔıkli :ni :kıjah̞

jumkınu stıxda :muh̞a: lıbına :ʔı lʔasa:sa:tı lati : tudaʕımu tıxa:dı lqara:rı lʔıklıni:ki: tahsi:n lʕıla:ʒı lfardi: jaqu:lu lbrufi:su:r ʒuwaki:m bi:lmu:nt mın mustaʃfa: lʒa :mıʕa lmustaqılah̞ wa mustaʃfa: di:l ma:r barʃalu:na ʔısba:nıja: ʔına lʕıla :ʒa lfardi: lʔamΘal lıkulı fard da:xıla masa:ra lʕıla:ʒ jumkınu ʔan jatahaqaq ʕan t̞ari:qı lʔaxðı fi : lʔıʕtıba:rııhtıja:ʒı kulı mari:d̞ın ʕala: hıdah̞ wa nah̞ʒa tarki:zıʕala : lmari:d̞ı laði: tama wad̞ʕuh̞u bıwa:sıt̞atı fari :qın mına lxubara:ʔı jad̞umu ʔat̞ıba :ʔa lmasa :lıkı lbawlıjatı wa lʔawra :mı mın ðawi: lxıbratı mın ʒami:ʕıʔanha :ʔıʔu:ru:ba: s̞umıma lıkaj juʕajına lʕawa :mıla lmuhadıdah̞ lati: janbaɤi: ʔaxðuh̞a: fi: lʔıʕtıba:r ʕında ıxtıja:rı lʕıla:ʒı lʔamΘalı lılmard̞a: kulun ʕala: hıdah̞(em es keı si: si:) anıð̞a :mu lʒadi:d jad̞aʕu fi: lʔıʕtıba:rı kulan mın mıqja:s murakaz ki:tri:nʒ atıðka:ri: lısarat̞a:nı wa tahli:lıʔansıʒatı lwaram wa ʕadad wa mawad̞ʕa ntıʃa:rı lwaram wa ʕumru lmari:d̞ wa ha:latıʔada :ʔı lmari:d̞ wa lʔamrad̞ı lmusa:hıbah̞ wa lʕawa :mılı lmurtabıt̞atı bılʕıla :ʒ wa bıstıxda:mı h̞a:ða lʔuslu:b qa :mat laʒnatun mına lxubara:ʔı bıstıʕrad̞ıʔahdaΘı lbaja:na:t lmuta:hatı lıtahdi:dı lmaʒmu:ʕa:t alfarʕıjah̞ mına lmard̞a: laði :na qar jastafi :du :na mına lʕıla:ʒı bısu:ra:fi:ni:b laði:na la: jatımu ʕa

:datan tamΘi:luḥum fi: taʒa:rɪbɪ lmarhalatɪΘa:lɪΘaḥ wa tama tahli:lu kulun mɪn nataʔɪʒɪ lmarhalatɪΘa :nɪjaḥ wa ΘalɪΘaḥ mɪna taʒa:rɪbɪ sari :rɪjaḥ wa bara:mɪʒɪ lwuṣulɪ lɪlmarḍa: lmuwasaʕaḥ wa tahli:la:tɪ lfarʕɪjaḥ wa dɪra:sa:tɪ lmara:kɪza lfardɪjaḥ. ḥuna:ka baja:na:t qawɪjaḥ tudaʕɪmu stɪxda:ma su:ra:fi:ni:b lɪṭaʔɪfatɪn mɪna lmaʒmu:ʕa:tɪ lfarʕɪjatɪ lɪlmarḍa: bɪman fi:ḥɪm lmu:sɪnu:na wa lmarḍa: laði:na juʃa:nu:na mɪna lfaʃalɪ lkalawi: wa tali:fɪ lkabɪd wa ladajḥɪm ɪntɪʃa :run lɪlwaram ʒajɪd ʔaw mutawasɪṭ wa rtifa:ʕɪ em es keɪ si: si: fi: ma: jaṣɪlu ʔɪla: Θala:ΘatɪʔaʕḍaʔɪI muxtalɪfatɪn mɪna lʒɪsmɪʒajɪd ʔaw mutawasɪṭ wa rtɪfa:ʕu ḍaɤtu damɪ lmuʕa:laʒ fatara:tu haja:tɪn ʔaṭwal fi: em es keɪ si: si: wa ladajḥɪm mɪqja:sun ka:na mutawasɪtun fatratɪ lhaja:tɪ mɪn du:nɪ taṭawurɪ eɪ a: si: si: es tahli:lu maʒmu :ʕatɪn farʕɪjatɪn mɪn baja:na:tɪ dɪra:satɪ lɪlmaraḍɪ mutama:Θɪl lɪlmarḍa: fawqa wa tahta sɪnu sabʕi:na ʕa:man Θala:Θatun wa ʕɪʃru :na faṣɪla tɪsʕaḥ muqa:bɪla sɪatun wa ʕɪʃru:na faṣɪla Θala:Θa ʔusbu:ʕan ʕala: tawa:li: kama : ka :nat tudaʕɪmu stɪxda:ma su:ra:fi:ni:b maʕa lmarḍa: laði:na la: juna :sɪbuḥum ʔaw la: jastaʒi:bu:na lɪlʕɪla:ʒɪ bɪsi:tu:ki:n ʕɪla:watan ʕala: qɪlatɪʕadadɪ lʔahda:Θɪ sɪlbɪjaḥ mɪna daraʒa:tɪΘala :Θaḥ wa ʔarbaʕaḥ lati: tuṣa:hɪbu ḥa :ða lʕɪla:ʒɪ fi : ha :latɪ kɪba :rɪ sɪnɪ muqa :ranatan bɪlʔadwɪjatɪ lmuwaʒaḥatɪ lʔuxra: kama: ʔaðḥara tahli:lun farʕɪjun ʔa:xara lɪdɪra:satɪ nafsuḥa: ʔana ɪstɪfa:data lmusɪni:na mɪna lʕɪla:ʒɪ ka :nat mutaqa:rɪbatan dɪra:sa:t qa:ʔɪmaḥ ataʒrubaḥ maʕa lʔaṣɤarɪ sɪnan maʕa majlɪn lɪwuʒu:dɪ fatratɪ baqa:ʔɪʔaṭwal mɪn du :nɪ taṭawur ḥalɪ lʕɪla:ʒu kaxaṭɪn ʔawalɪn bɪ su :ra :fi :ni :b swɪtʃ fi : lʕɪla :ʒɪ lmutaʕa:qɪbsajatɪmu lʒawabu ʕala: ḥa:ða suʔa:lɪʕan ṭari:qɪ lmarhalatɪΘa:lɪΘatɪ lʕaʃwa:ʔɪjatɪ wa lmaftu:hatɪ lɪtaʒrubatɪ dawlɪjaḥ ḥa:ðɪḥɪ dɪra:satu badaʔatɪ fi: jana:jar ka:nu:n aΘa:nɪʔalfajnɪ wa tɪsʕaḥ wa juxaṭaṭu laḥa: lɪstɪmra:ru hata: ʕa:ma ʔalfajnɪ wa ʔɪΘna: ʕaʃar wa taʃmulu xamsatun wa ʔarbaʕu:na mari:ḍan ʔaḥamɪ xaṣaʔɪṣa ʔɪdmaʒɪḥɪm fi: taʒrubatɪ ḥɪja ʔana marḍa: saraṭanu xala:ja: lkɪla: lmaṣhu:bɪ bɪntɪʃa:rɪ lɪlwaramɪ wa la: jaṣlahu:na lɪlʕɪla:ʒɪ bɪsi:tu:ki:na:t laði :na jaku :nu dawa:ʔu bɪnɪsbatɪʔɪlajḥɪm ḥuwa lxaṭu lʔawalu lɪlʕɪla:ʒɪ ðɪra :ʕu lʔawalu jatɪmu ʔɪʕṭa:ʔuḥum su:ra:fi:ni:b wa baʕda waqfɪ lʕɪla:ʒɪ bɪsababɪ taṭawurɪ lmaraḍɪʔaw ðuḥu:rɪʔaʕraḍɪn ʒa :nɪbɪjatɪn jatɪmu ʔɪʕṭa:ʔuḥum sni:ti:ni:b wa lʕaksu ṣahi :h bɪnɪsbatɪ lɪðɪra:ʕɪΘa:nɪjaḥ. alḥadafu raʔi:si: ḥuwa taqji:mu ma: ʔɪða: ka:natɪ fatratu lbaqa:ʔɪ xa :lɪjatan mɪn taqadumɪ lmaraḍɪ mɪn waqtɪ lɪxtɪja:rɪ lʕaʃwa:ʔi: hata: taṭawurɪ lmaraḍɪʔawɪ lwafa:t wa ʔɪhda: nɪqa:ṭɪ lati: tartakɪzu ʕalajḥa: dɪra:saḥ ḥɪja tahli :lu sɪmjaḥʕaḍalatu lqalbɪ lati: tatɪmu ʕan ṭari:qɪ rasmɪ lqalbɪ wa tahli:lɪ muʔaʃɪr ḥubu:ṭɪ lqalbɪ

duktu:r ʕabdu lhafi:ð jahja: xu:ʒaḥʃarqɪ lʔawṣat

Translation of Arabic Text

Kidney Cancer

New Treating Styles

Effective Cures Give Patients Big Hope for Longevity

Kidney cancer is one among lethal health problems; it is therefore the second fundamental cause of death in the world. Statistics regarding genital and urinary cancer systems states that there is annual increase in kidney cancer patients in the Arab world which goes on a par with the world increase. It was found that kidney cancer cases have increased by **52%** during the period between **1983** and **2002**, that is, from **7.1** to **10.8** case for each **100.000** persons. Simply put, the death rates have increased mainly among those who suffer from tumours of greater size than 7 cm. It rose then from **1.2 to 3.2** to each **100.000** persons; thus many statistics also pointed out to the rise of kidney patients in the Arab world. For instance, the Saudi statistic register of tumours shows that the rate of kidney cancer has reached 200 cases per year more than it is in Mecca, Ryad and the Eastern Region. In order to discuss the new medical and scientific update of kidney cancer, the Saudi Association of Urology, with the collaboration of Saudi Association of Tumours, held a medical conference in Sharm El Sheikh from February **10–12, 2010**.

During his speech, the president of conference, Dr. Ashraf Abou Sacra, an adviser of surgical urology (medical town of the king Abdulaziz in Jeddah), stated that the fundamental aim of the conference is to raise the level of awareness among physicians and consequently patients and to give attention to consolidate; prevention is better than cure, and the importance and seriousness of kidney cancer disease particularly after the rise of the disease rate in the whole world is referred. Doctor Abu-Sacra from his side has confirmed people who have a disease and family history regarding kidney cancer to tell their doctors and do the necessary tests repeatedly because doing a positive role towards health preservation increases the chances to detect the disease earlier. Abu-Samra added that there are many options available to treat renal cell carcimona and can give more than one type of treatment at the same time depending on the pathological stage of cancer. There is radical nephrectomy, lacroscopic nephrectomy, the removal of widespread impunities, radiation therapy, bioremediation and target therapy; the latter is a glimmer of hope to these patients because it has proven its effectiveness in treating this kind of cancer like nexavar and other similar drugs from the same type.

Surgery is the first means in treating renal cell carcimona and has the possibility of healing, but medium cancer or high-risk cancer often comes back after surgery in the 35 × 65% of cases; this reality is restricted by the following study which contained 1671 patients who suffer cell kidney cancer of the type 'clear cell' topical, and in one side of the body, they underwent a surgery of radical removal of kidney (Leibovich BC et al. Cancer 2003: 97:1663–71). There has been a spread of the disease in 479 patients during 1.3 years on the average. The average period of survival without the spread of tumour was 86.9% in 1 year and 77.8% in 3 years and 67.1% in 10 years. New hope Professor Peter Molders from medical center in Nyiemgin at Radwood University in the Netherlands says that the groups selected from the patients have a low probability of disease progression after kidney removal. The adjuvant treatment (after surgery) by radiotherapy, hormonetherapy, chemotherapy and cytokines showed poor efficiency in treating kidney cancer cells; nevertheless there is a new hope. He adds 'the targeted drugs, such as, tyrosine kinase inhibitors via oral way, are perfect and efficient in the treatment of kidney cancer cells and fit the long-term treatment, these drugs can provide choices to treatment after surgery in future. As for the next stage regarding the clinical trials with these drugs, it is currently under way including the assured experience which means the use of sorafenib as adjunctive treatment in kidney cancer case of favourable result and experience source which means to compare sorafenib with placebo in patients who suffer from cancer cell kidney after initial eradication. Doctor Molders pointed out that there is still a clear clinical need for treatment assistance. The best solution, says Professor Yorgen Jeshund from medical centre Rickts der Isar at Munich University of Arts in Germany, is that currently there are a number of strategies available and on the way of development, in order to bettering treatment outcomes. This may include determining the optimal dose, treatment side effects and mixing and sequencing optimal treatments. He adds that the use of oriented drugs may enhance the clinical activity by targeting signals on multiple levels; whereas the therapy oriented by the immune therapy is under investigation. What about alternative therapy? Is it better? Dr. Jesund says that current evidence suggests that the use of growth inner wall factor inhibitors of blood vessels can be useful without hampering the following response to inhibitors of mammalian target of rapamycin (MTOR); therefore the maximum period to achieve clinical benefit can be made using two (VEGF) inhibitors, prior to transforming the way of work to inhibitors of targeted therapies. As for the role of targeted therapies in treating kidney cancer cells, it should be noticed that the way in treating kidney cancer cells has changed dramatically in the beginning of the twenty-first century, due to insightful knowledge of the science of molecular biology with regard to tumours and new treatment options that are available. And since kidney cancer cells resist chemotherapy significantly, cytokines have been used for advanced cases treatment; nevertheless the latter is effective only in a limited number of patients and is unsuitable for the majority of patients due to toxicity. The first type of targeted therapies designed to inhibit tyrosine kinase receptors, which are believed to be important for the growth of tumour-feeding blood vessels, was sorafenib. It has given clearance for treating cancer kidney cells in 2005 in United States and in EU in 2006, in addition to tyrosine kinase inhibitor sunitinib in 2006 and targeted inhibitor of the mammalian Sirolimus Rapamycin, vascular wall growth factor inhibitor and interferon for cancer kidney cells in 2007. Professor Zaya Kirkala from the Izmir University in Turkey, Faculty of Medicine, was asked how to choose one thing among many choices. His response was that every patient is different from the other so one treatment cannot fit all the patients. And he gave an example about the elderly patients. On the basis of that fact, the International Association for Elderly Cancer recommends the following: ‘When taking into account the most appropriate medicines for use with a particular patient, we should take into account the toxicological data per each treatment on a uniform basis and the presence of another specific disease’. Thus, all data – from non-randomized studies, sub-analyzes, extended patient access programmes, and subprogrammes analysis, and expanded programmes for the access to patients, and retrospective studies, and case studies, and clinical experience can be used to build foundations that back up clinical decision-making for the improvement of individual therapy.

Professor Joaquim Belmont from the Independent University Hospital and Hospital del Mar in Barcelona, Spain, says: ‘The optimal individual therapy for each individual within the course of treatment can be achieved by taking into account the need for each patient individually, and the way regarding the focus on the patient, which was made by a group of experts including urologists and oncholo-gists with expertise from all over Europe’. It was designed to appoint specific fac-tors that should be taken into consideration when choosing the optimal treatment for individual patients. The new system takes into consideration both of the standard center Memorial Sloan Kettering Memorial Cancer (MSKCC) and analysis of tumour tissue, including the number and positions of the spread tumour, the age of the patient, the status of the performance of the patient, other diseases and factors related to treatment. Using this method, a panel of experts reviewed the latest avail-able data to identify the subgroups of patients who may benefit from the treatment with ‘sorafenib’, which are not usually represented in trials of phase III. Results of phase II and phase III were analysed from clinical trials, programmes of expanded access for patients, subanalysis and individual centre studies. There is strong data which support the use of ‘sorafenib’ to a range of subgroups of patients, including the elderly and patients who suffer from kidney failure and cirrhosis of the liver and have the spread of tumour in up to three different organs of the body and a good or average measure of MSKCC and a high blood pressure.

‘A longer lifespan’ in the analysis of a subset of the study data, the average lifes-pan without the development of the disease similar to patients above and below the age of 70 (23.9 versus 26.3 weeks, respectively) is shown. It supported also the use of sorafenib with patients whose body rejects it or did not respond to treatment with cytokine, in addition to the small number of negative events of grades 3–4 which accompany this treatment in the case of elderly compared to other targeted med-cines. Another sub-analysis of the study itself showed that elderly benefit from treatment was similar to younger patients, with the tendency for a longer period of survival without the development.

Is treatment as a first line with sorafenib more useful than **sunitinib** in successive treatment? This question will be answered through the random third phase, opened to the international experience ‘Switch’. This study began in January 2009 and planned to continue until 2012. It includes 540 patients where the most characteris-tics of their integration in the experiment are to suffer from kidney cancer cells coupled with the spread of the tumour and do not respond to treatment with cyto-kine, and medicine is the first line of treatment. The first group is given sorafenib, and after stopping treatment because of disease progression or appearance of some effects, they are given sunitinib and vice versa for the second group. The fundamen-tal objective is to assess whether the period of survival is free of disease progression from the time of random selection until disease progression or death. One of the focal points of the study is to analyze the toxic of its heart muscle, which is made by drawing a heart and analyzing the decline in heart failure.

Yahia abdul Hafeez Khoja Source. Middle East

Referential Samples

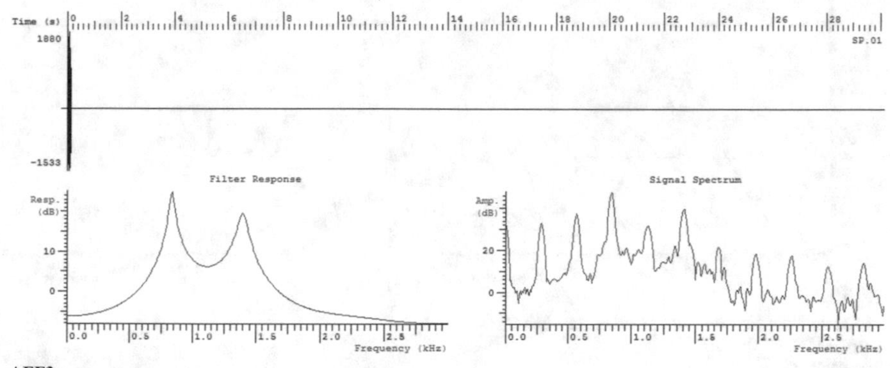

AFF3
F1 835Hz F2 1399Hz
Amp: 0.1603dB
Per: 0.0001667

....

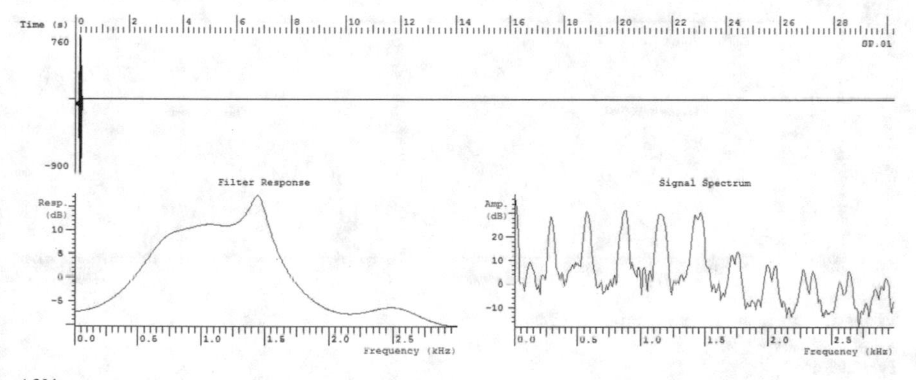

Afd4
F1 717Hz F2 1058Hz F3 1454Hz F4 2520Hz
Amp: 0.2603 dB
Per: 0.0001667

....

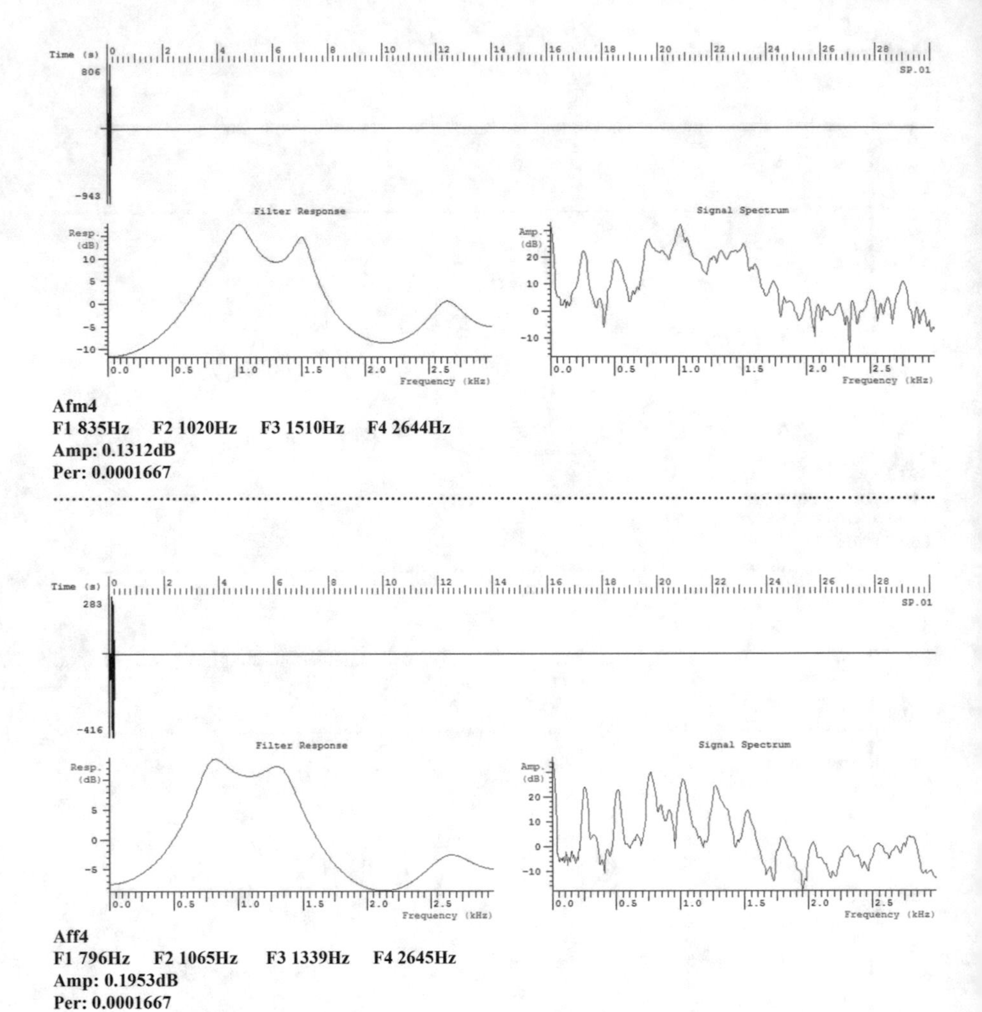

Afm4
F1 835Hz F2 1020Hz F3 1510Hz F4 2644Hz
Amp: 0.1312dB
Per: 0.0001667

………

Aff4
F1 796Hz F2 1065Hz F3 1339Hz F4 2645Hz
Amp: 0.1953dB
Per: 0.0001667

………

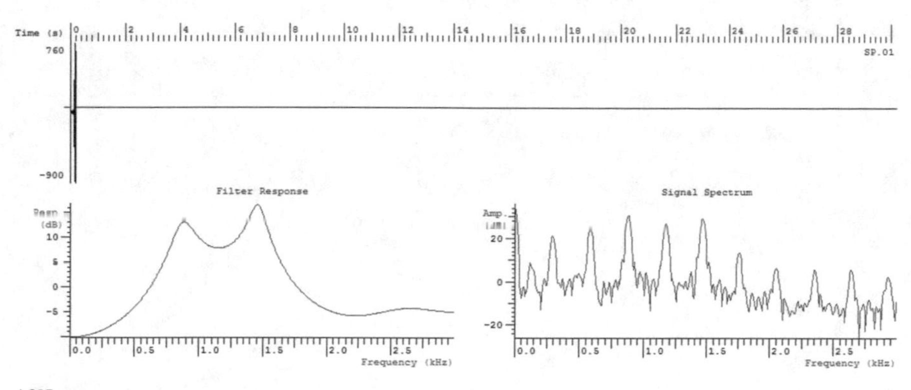

Afd5
F1 875Hz F2 1462Hz F3 2625Hz
Amp: 0. 1962dB
Per: 0.0001667

....

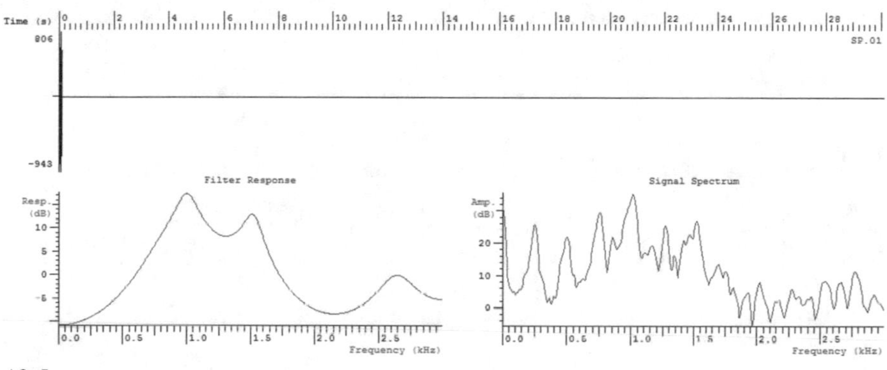

Afm5
F1 813Hz F2 1002Hz F3 1520Hz F4 2638Hz
Amp: 0.1286dB
Per: 0.0001667

....

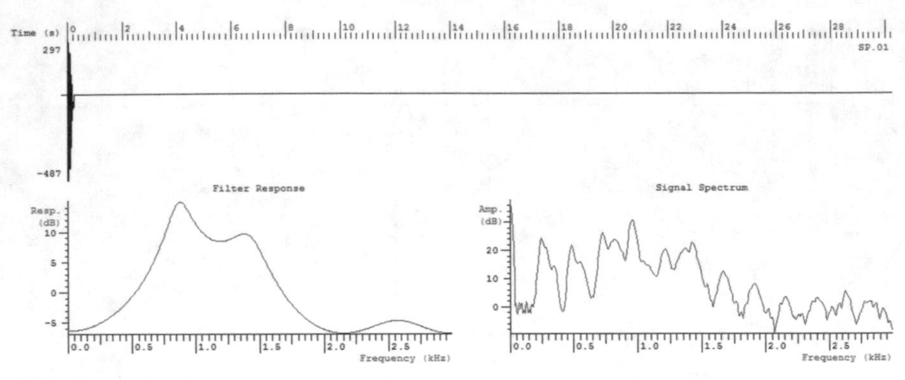

Aff5
F1 871Hz F2 1419Hz F3 2571Hz
Amp: 0.2577dB
Per: 0.0001667

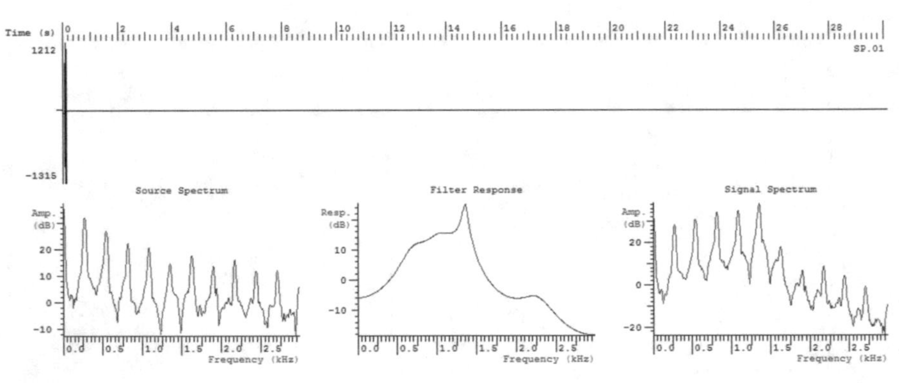

Agd1
F1 683Hz F2 1038Hz F3 1359Hz F4 2269H
Amp: 0.1609dB
Per: 0.0001667

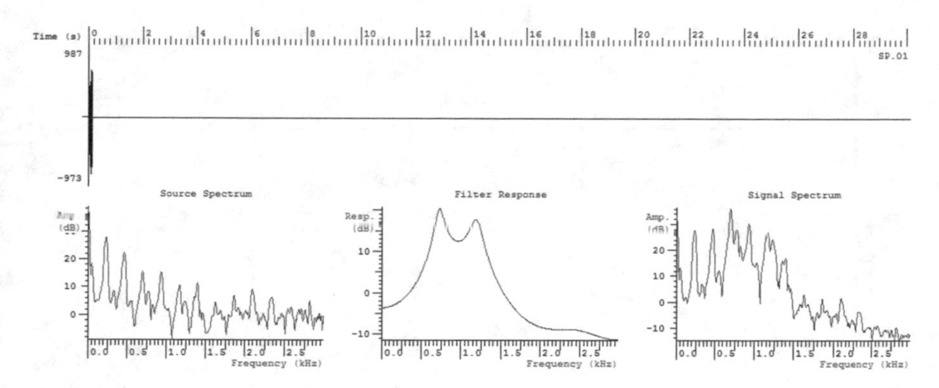

Agm1
F1 721Hz F2 1195Hz F3 2480Hz
Amp: 0.1600
Per: 0.0001667

....

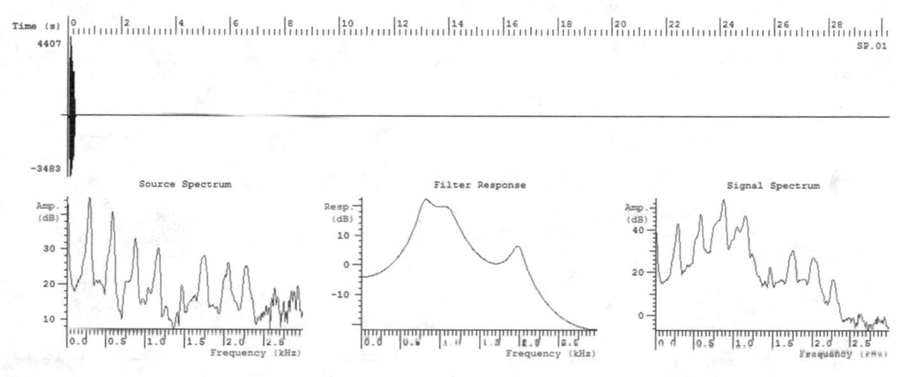

Agf1
F1 801Hz F2 1098Hz F3 1986Hz
Amp: 0.2894dB
Per: 0.0001667

....

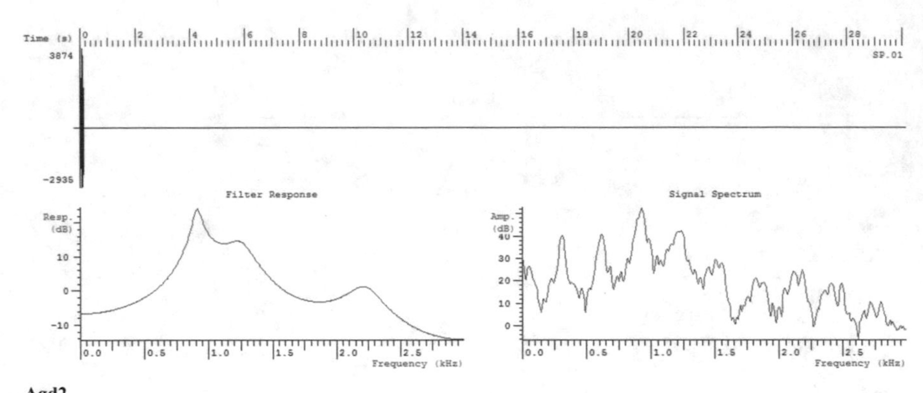

Agd2
F1 911Hz F2 1257Hz F3 2225Hz
Amp: 0.1563dB
Per: 0.0001667

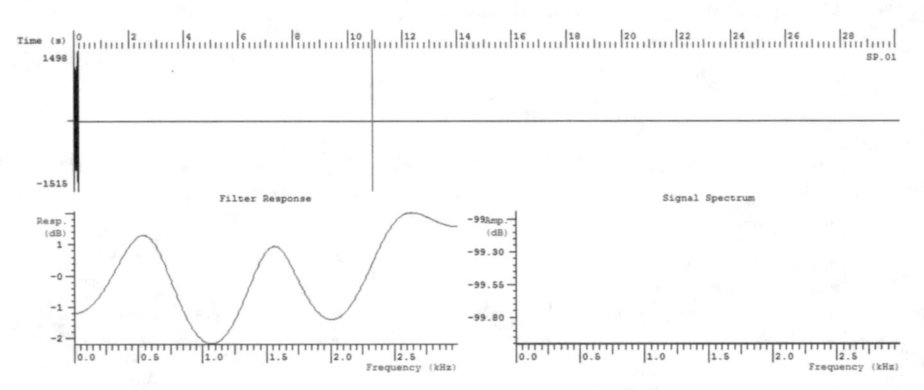

Agm2
F1 564Hz F2 1550Hz F3 2520Hz
Amp: 0.1872dB
Per: 0.0001667

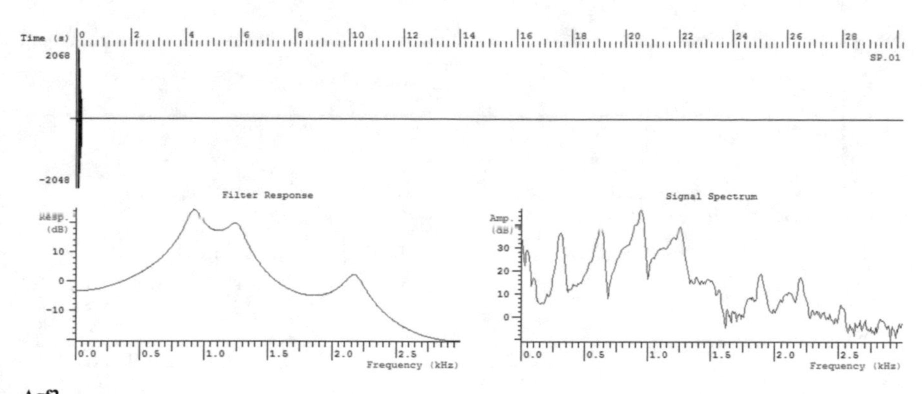

Agf2
F1 646Hz F2 918Hz F3 1255Hz F4 2180Hz
Amp: 0.2214dB
Per: 0.0001667

....

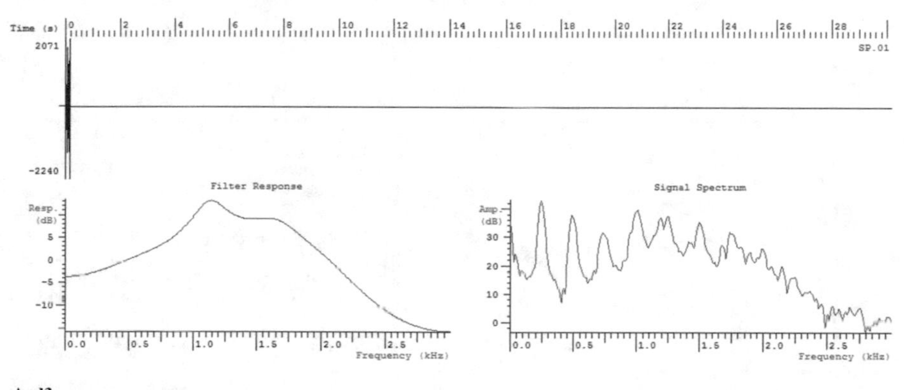

Agd3
F1 537Hz F2 1127Hz F3 1654Hz F4 2051Hz
Amp: 0.1878dB
Per: 0.0001667

....

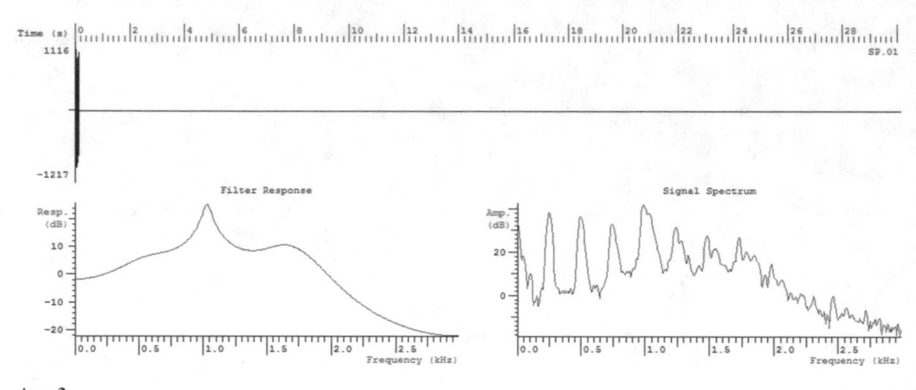

Agm3
F1 541Hz F2 1032Hz F3 1658Hz F4 1818Hz
Amp: 0.1560dB
Per: 0.0001667

....

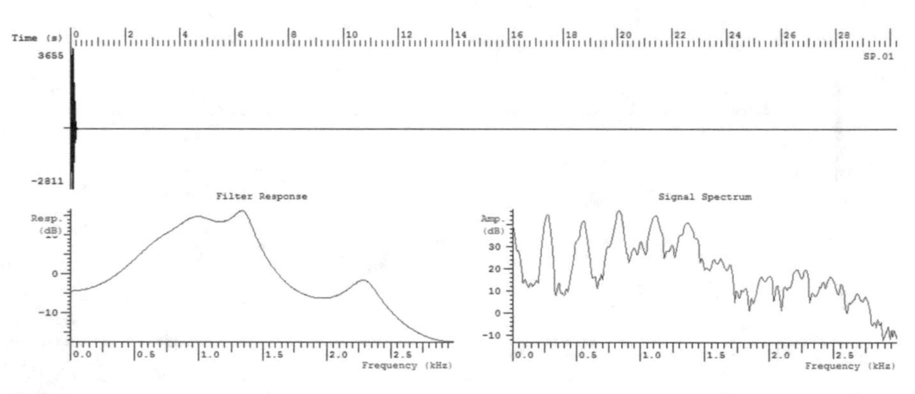

Agf3
F1 653Hz F2 983Hz F3 1352Hz F4 2299Hz
Amp: 0.2275dB
Per: 0.0001667

....

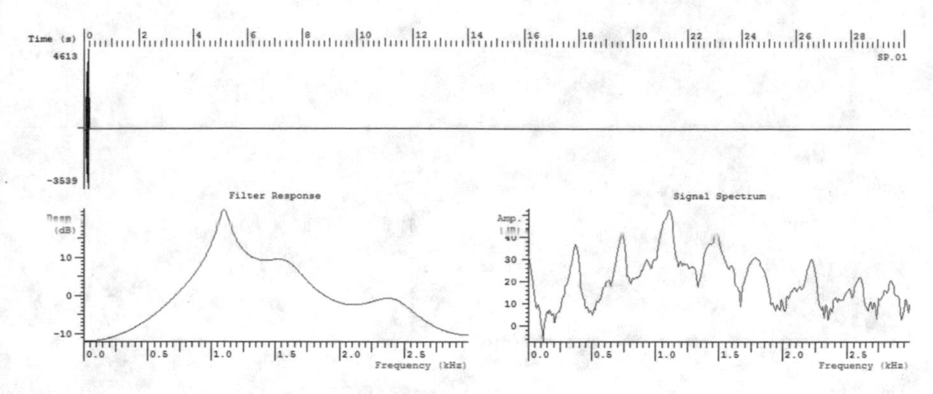

Agd4
F1 853Hz F2 1096Hz F3 1584Hz F4 2411Hz
Amp: 0.1953 dB
Per: 0.0001667

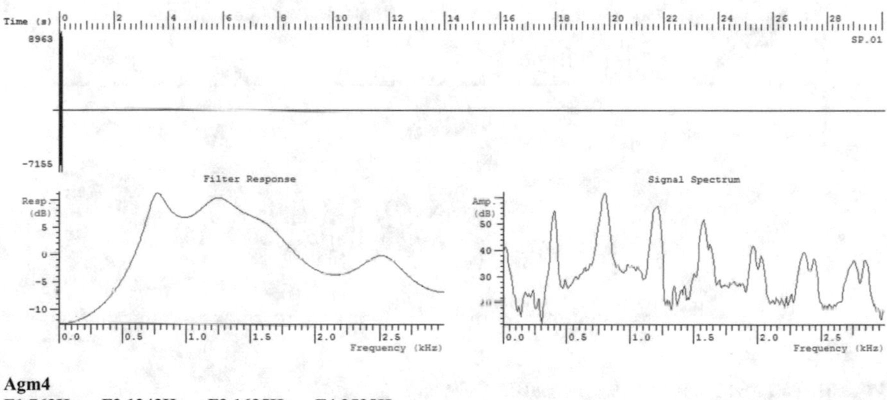

Agm4
F1 763Hz F2 1243Hz F3 1635Hz F4 2528Hz
Amp: 0.1291 dB
Per: 0.0001667

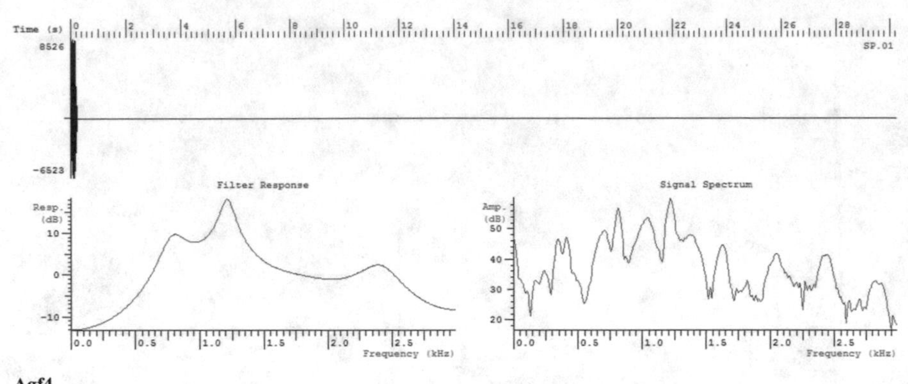

Agf4
F1 785Hz F2 1215Hz F3 1782Hz F4 2416Hz
Amp: 0.2539dB
Per: 0.0001667

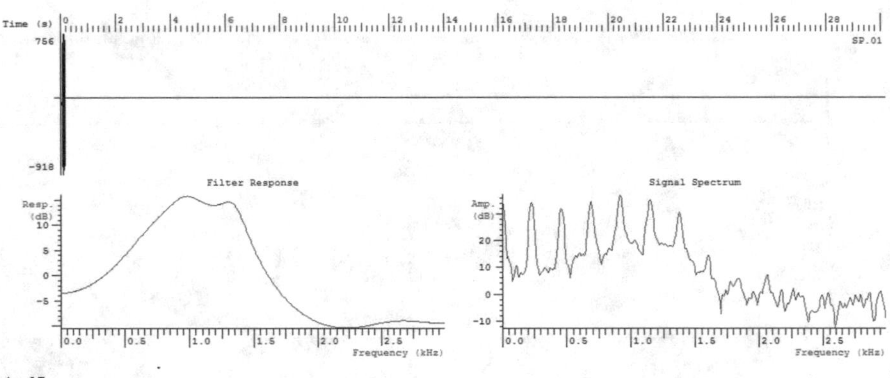

Agd5
F1 688Hz F2 972Hz F3 1347Hz F4 2631Hz
Amp: 0.1915dB
Per

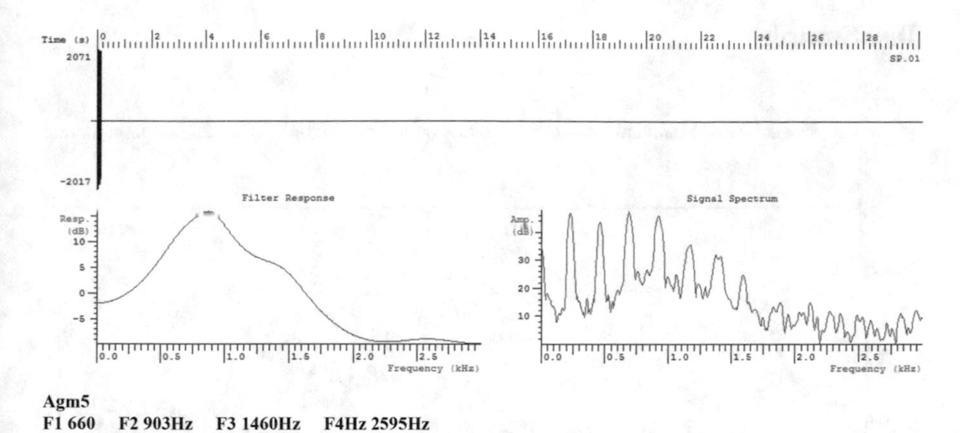

Agm5
F1 660 F2 903Hz F3 1460Hz F4Hz 2595Hz
Amp: 0.1606dB
per

....

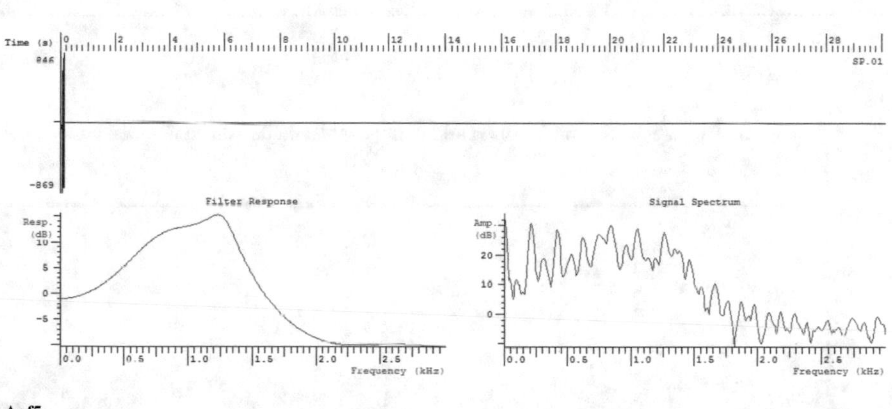

Agf5
F1 743Hz F2 905Hz F3 1271Hz F4 2605Hz
Amp: 0. 1606dB
Per: 0.0001667

....

Test Samples

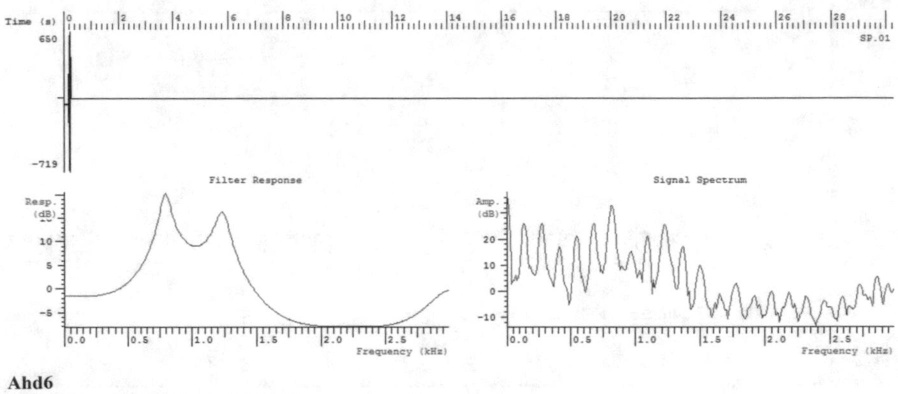

Ahd6
F1 796Hz F2 1239Hz F3 2206Hz
Amp: 0.2603dB
Per: 0.0001667

....

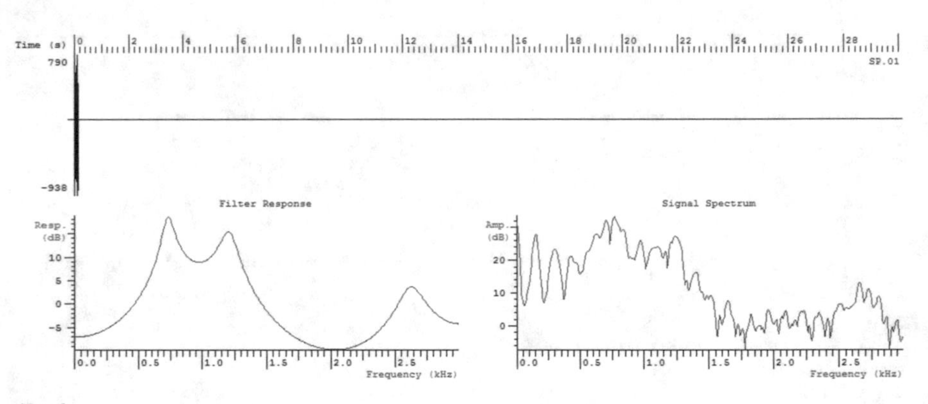

Ahm6
F1 733Hz F2 1213Hz F3 2627Hz
Amp: 0.1320dB
Per: 0.0001667

....

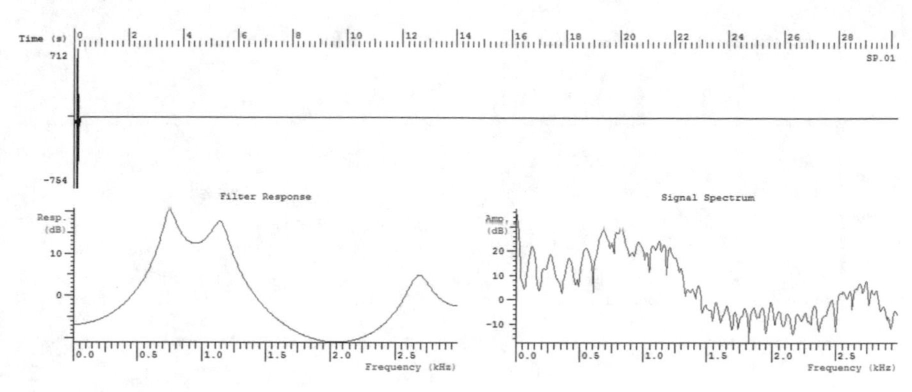

Ahf6
F1 747Hz F2 1143Hz F3 2684Hz
Amp: 0.2278 dB
Per: 0.0001667

....

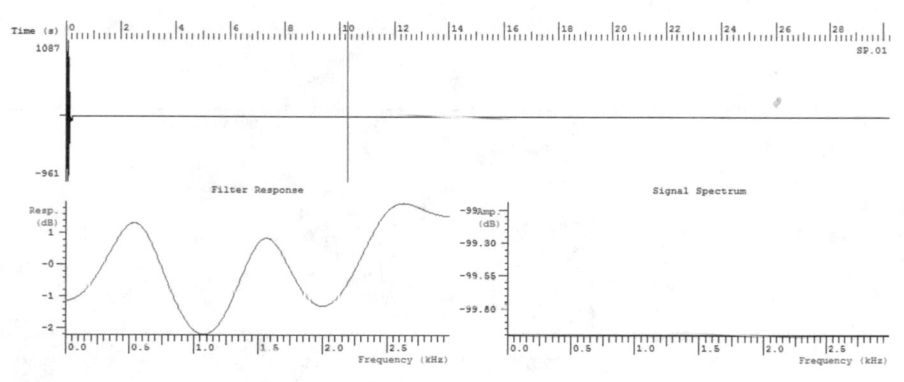

Ahf7
F1 565Hz F2 1552Hz F3 2505Hz
Amp: 0.2275dB
Per: 0.0001667

....

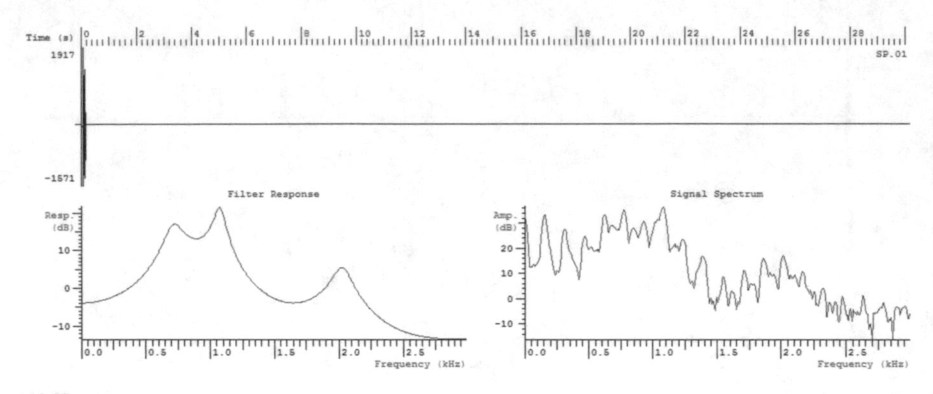

Ahd7
F1 714Hz F2 1075Hz F3 2028H
Amp: 1638dB
Per: 0.0001667

....

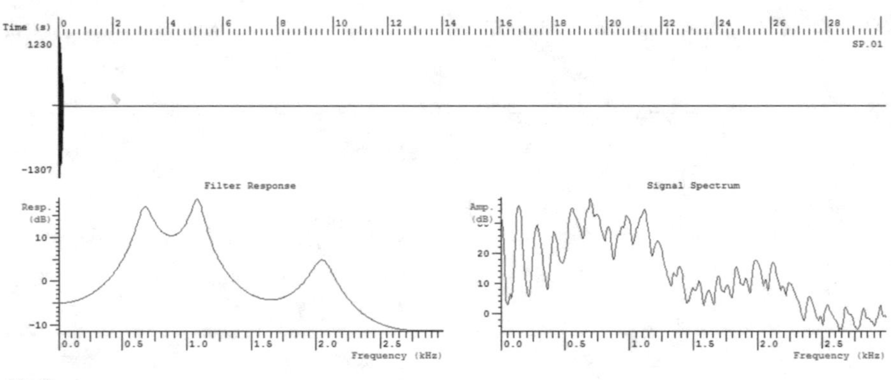

Ahm7
F1 672Hz F2 1085Hz F3 2054Hz
Amp: 0.1652
Per: 0.0001667

....

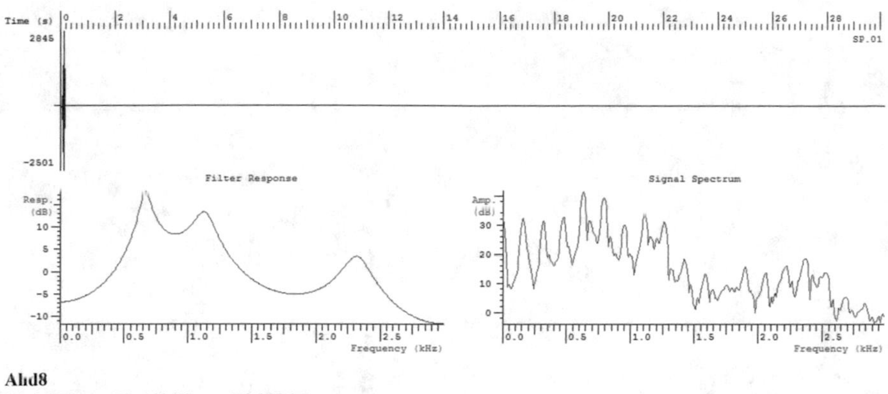

Ahd8
F1 663Hz F2 1134Hz F3 2326Hz
Amp: 0.1641dB
Per

....

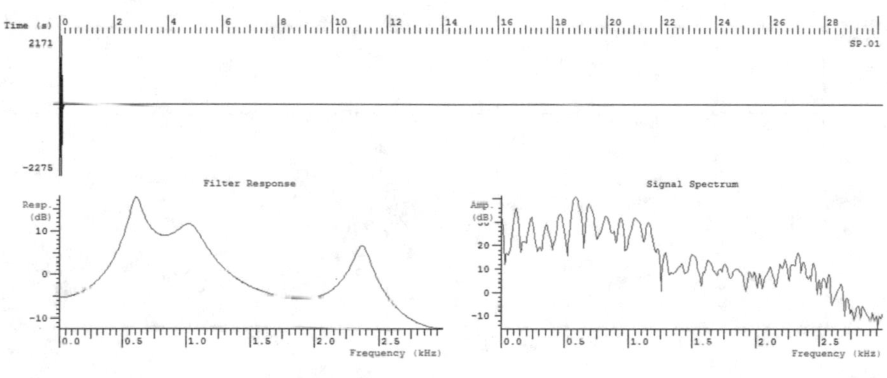

Ahf8
F1 602Hz F2 1037Hz F3 2374Hz
Amp: 0.1353dB
Per

....

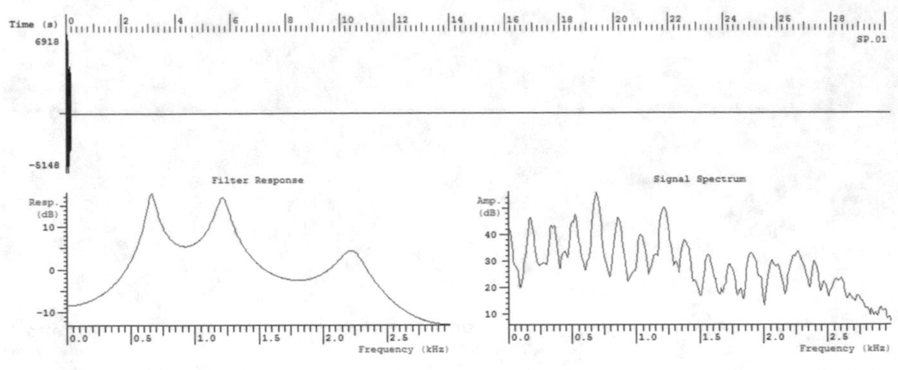

Ahm8
F1 663Hz F2 1219Hz F3 2234Hz
Amp : 0.1632
Per:

....

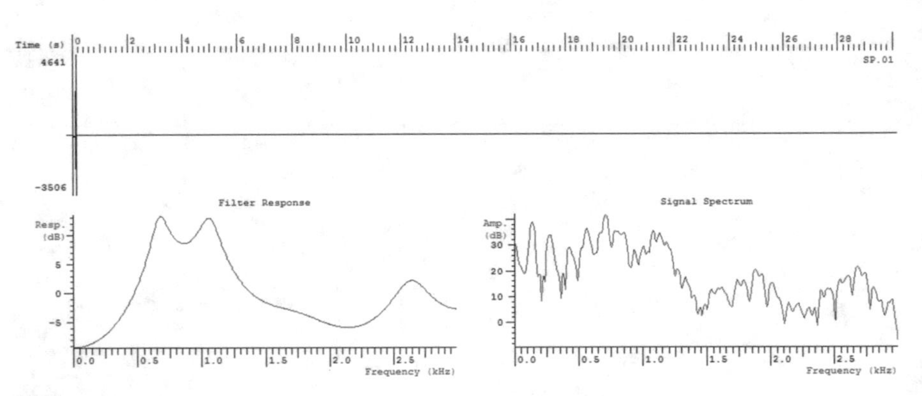

Ahd9
F670Hz F1068Hz F1750Hz F2641Hz
Amp: 0.1600dB
Per: 0.0001667

....

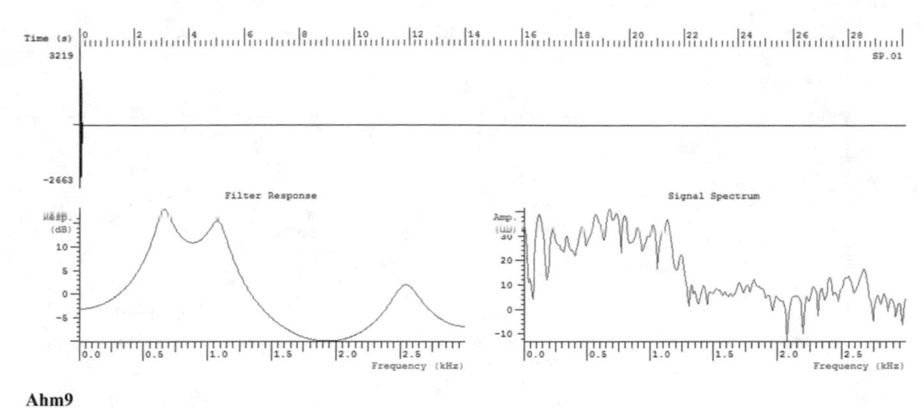

Ahm9
F1 661Hz F2 1088Hz F3 2543Hz
Amp: 0.0998dB
Per: 0.0001667

……………………………………………………………………………………………………

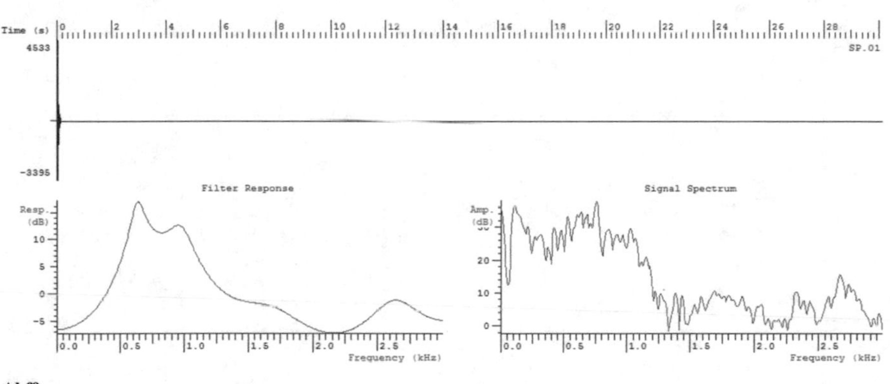

Ahf9
F1 632Hz F2 976Hz F3 1689Hz F4 2634Hz
Amp: 0.1312dB
Per: 0.0001667

……………………………………………………………………………………………………

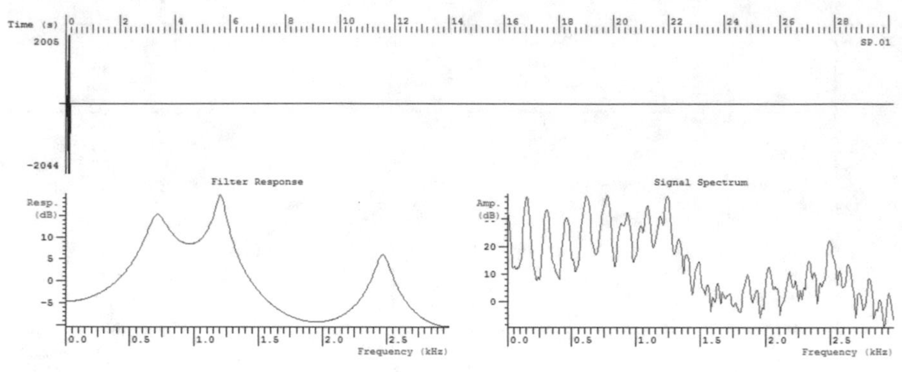

Ahd10
F1 714Hz F2 1209Hz F3 2474Hz
Int : 0.1635dB
Per : 0.0001667

....

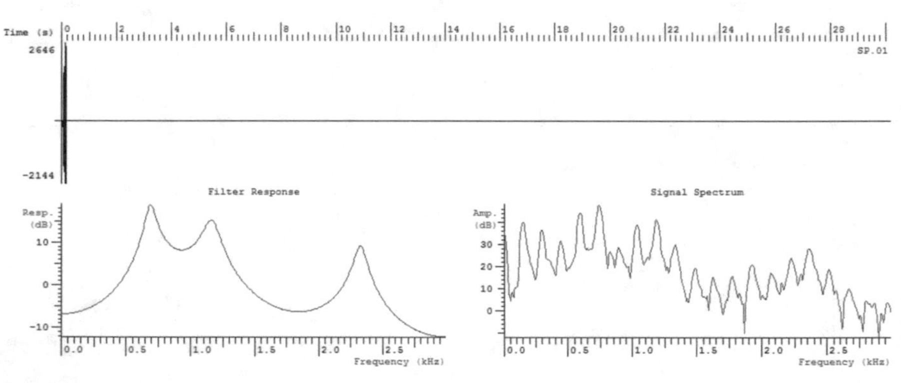

Ahm10
F1 691Hz F2 1167Hz F3 2329Hz
Int : 0.1910dB
Per : 0.0001667

....

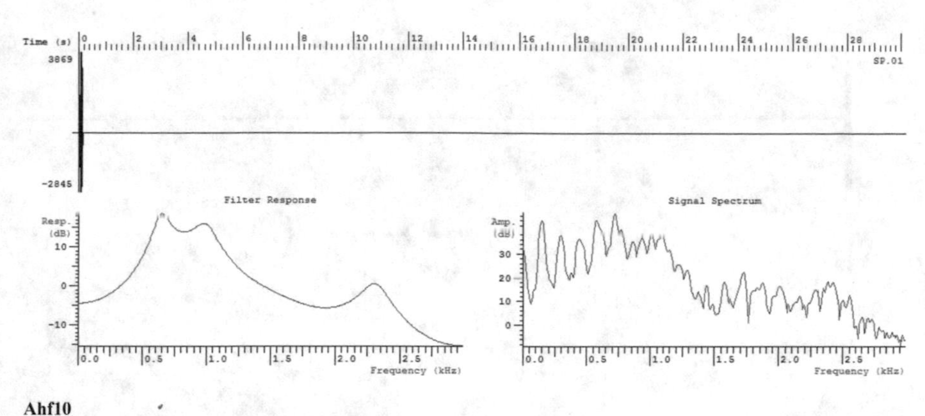

Ahf10
F1 642Hz F2 994Hz F3 1498Hz F4 2315Hz
Int : 0.1626dB
Per : 0.0001667

....

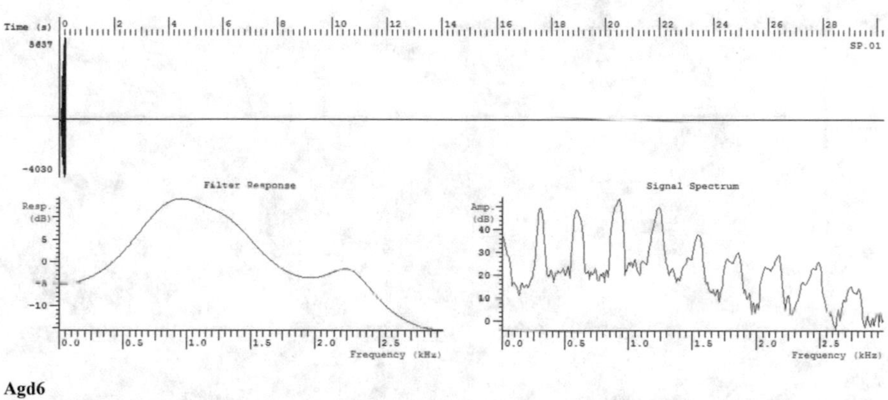

Agd6
F1 791dB F2 983dB F3 1353Hz F4 2274Hz
Int : 0.2545
Per : 0.0001667

....

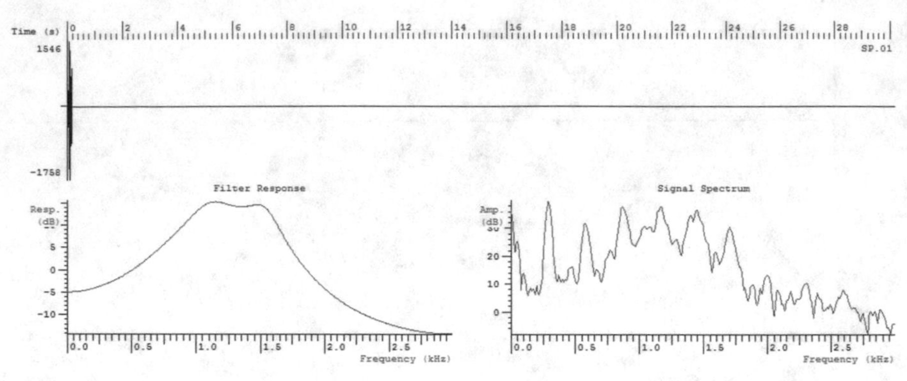

Agm6
F1 830Hz F2 1108Hz F3 1534hz
Int: 0.1907dB
Per: 0.0001667

……………………………………………………………………………………………………

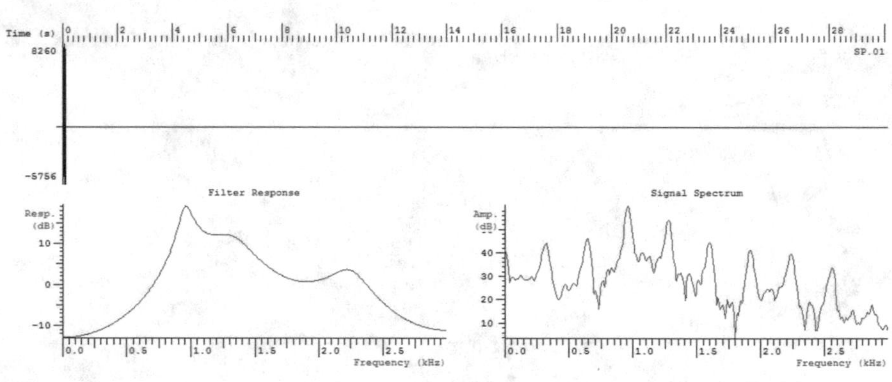

Agf6
F1 956Hz F2 1330Hz F3 2249Hz
Int: 0.1291dB
Per: 0.0001667

……………………………………………………………………………………………………

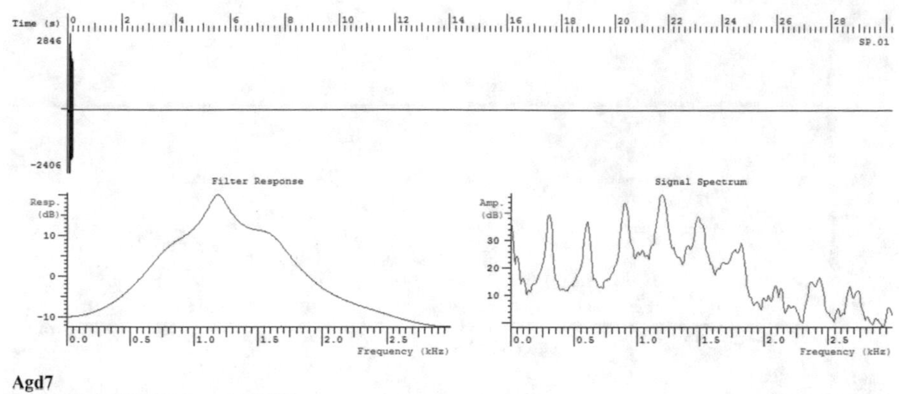

Agd7
F1 795Hz F2 1184Hz F3 1601Hz
Int: 0.2231 dB
Per: 0.0001667

....

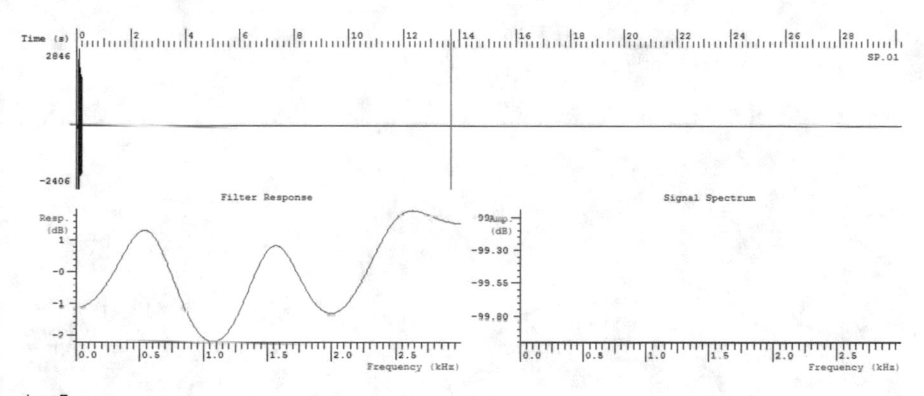

Agm7
F1 565Hz F2 1552Hz F3 2505Hz
Int: 02228dB
Per: 0.0001667

....

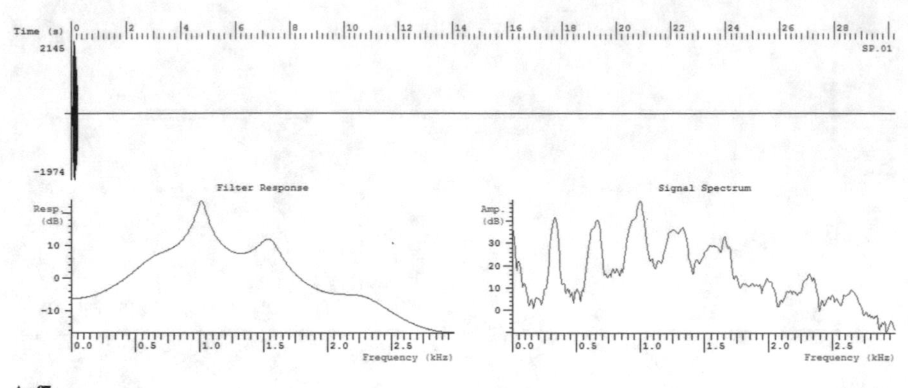

Agf7
F1 634Hz F2 1015Hz F3 1547Hz F4 2295Hz
Int: 0.2545dB
Per: 0.0001667

....

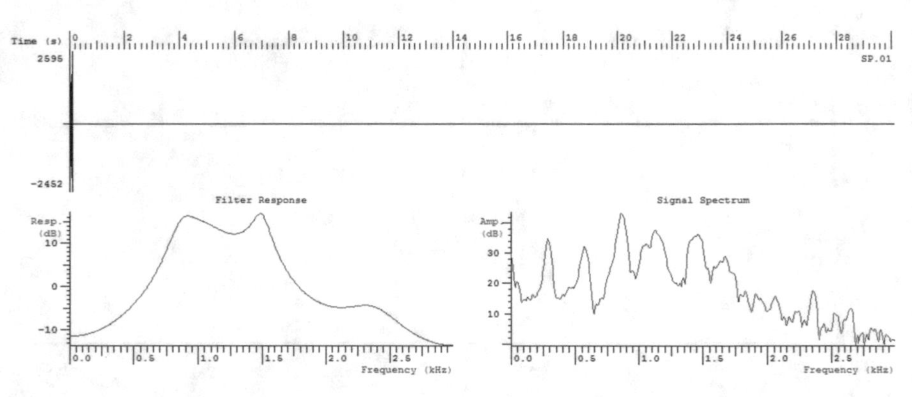

Agd8
F1 873Hz F2 1074Hz F3 1497Hz F4 2377Hz
Int : 0.1289dB
Per : 0.0001667

....

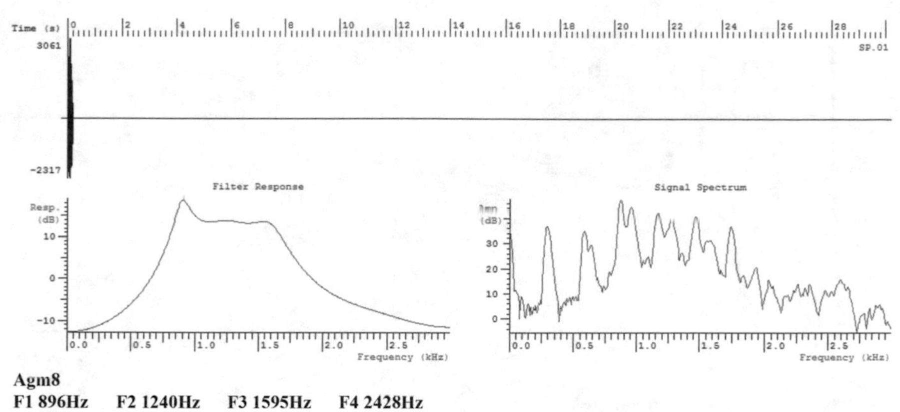

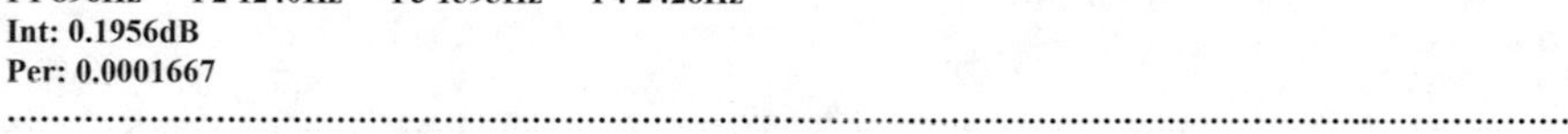

Agm8
F1 896Hz F2 1240Hz F3 1595Hz F4 2428Hz
Int: 0.1956dB
Per: 0.0001667

....

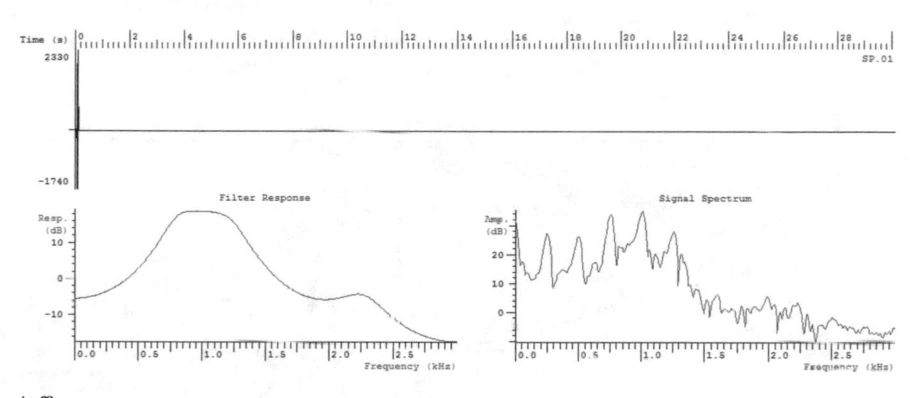

Agf8
F1 801Hz F2 1028Hz F3 1235Hz F4 2269Hz
Int: 0.1312dB
Per: 0.0001667

....

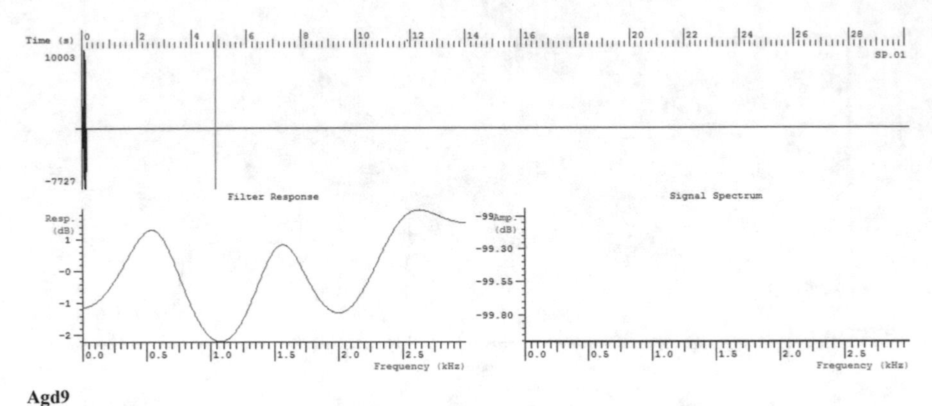

Agd9
F1 565Hz F2 1552Hz F3 2505Hz
Int: 0.1635dB
Per: 0.0001667

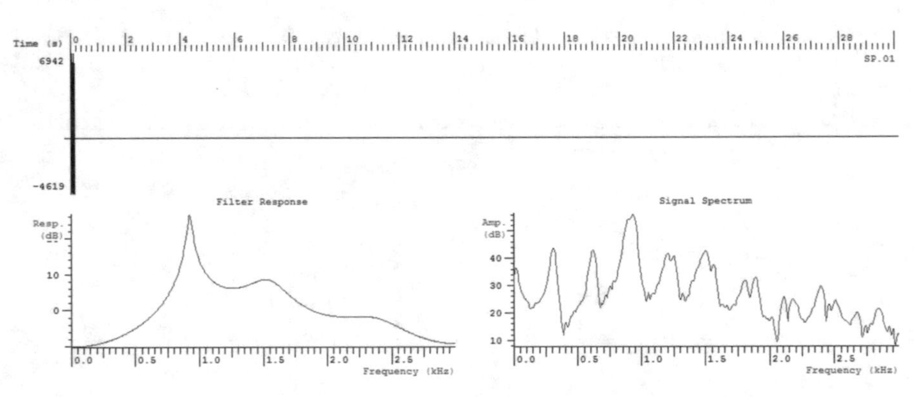

Agm9
F1 925Hz F1 1538Hz F3 2373Hz
Int: 0.1317dB
Per: 0.0001667

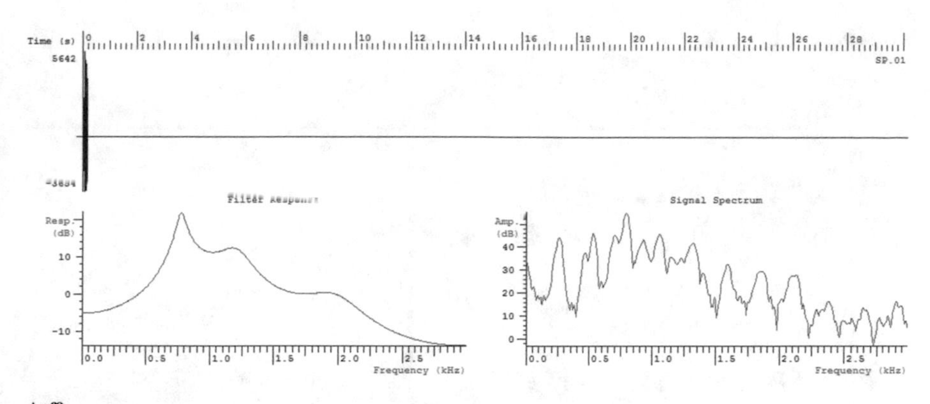

Agf9
F1 776Hz F2 1205Hz F3 1979Hz
Int: 0.1915dB
Per: 0.0001667

....

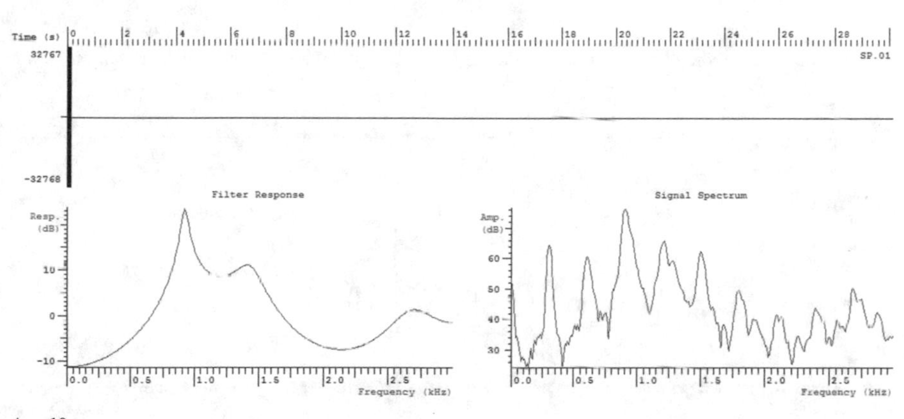

Agm10
F1 918Hz F2 1434Hz F3 2684Hz
Int: 0.1439dB
Per: 0.0001667

....

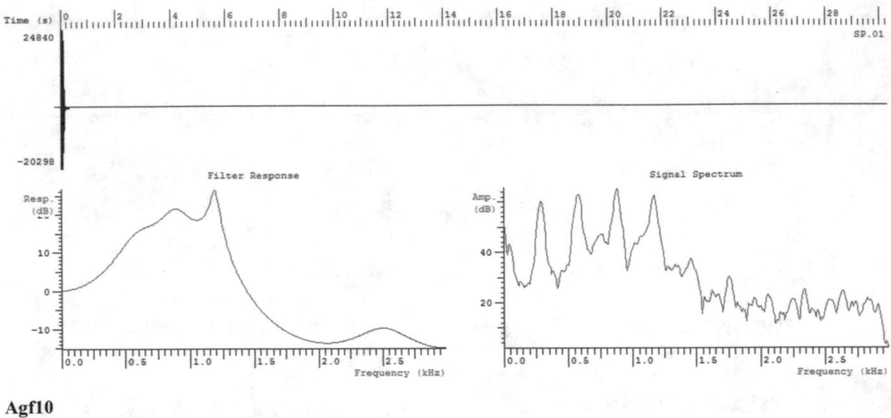

Agf10
F1 581Hz F2 887Hz F3 1191Hz F4 2513Hz
Int: 0.1919
Per: 0.0001667

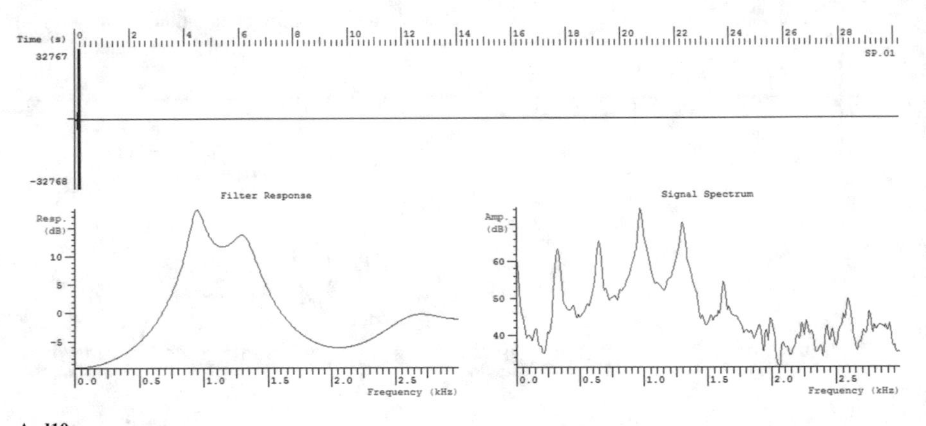

Agd10
F1 947Hz F2 1318Hz F3 2645Hz
Int: 0.2205dB
Per: 0.0001667

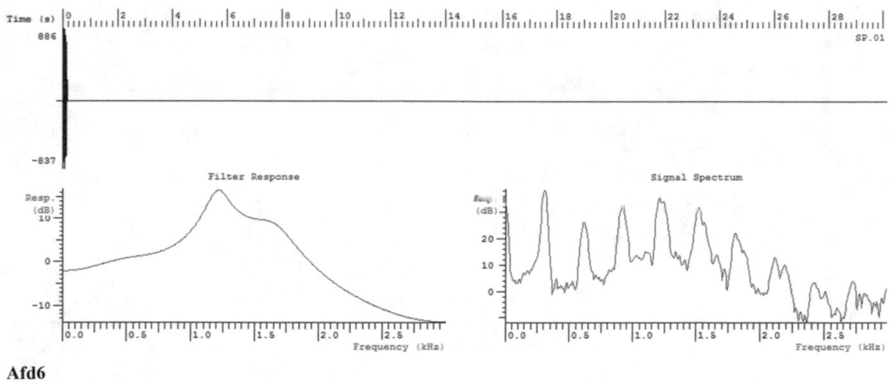

Afd6
F1 466Hz F2 1210Hz F3 1659Hz
Int: 0.1635dB
Per: 0.0001667

……………………………………………………………………………………………

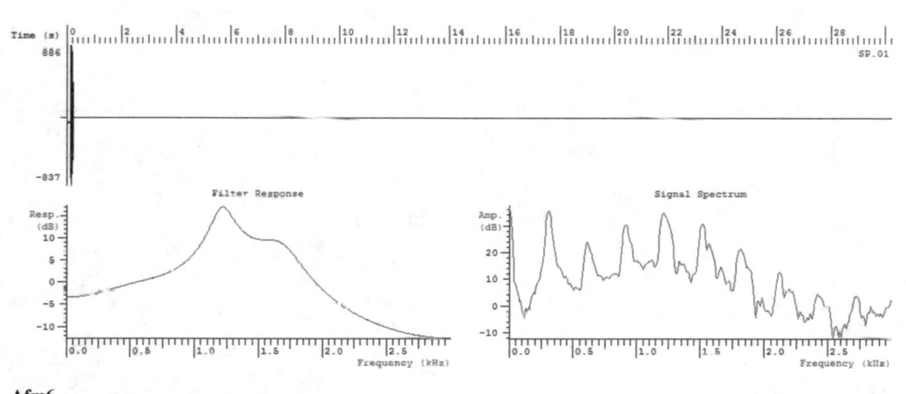

Afm6
F1 502Hz F2 1217Hz F3 1681Hz
Int: 0.2280dB
Per: 0.0001667

……………………………………………………………………………………………

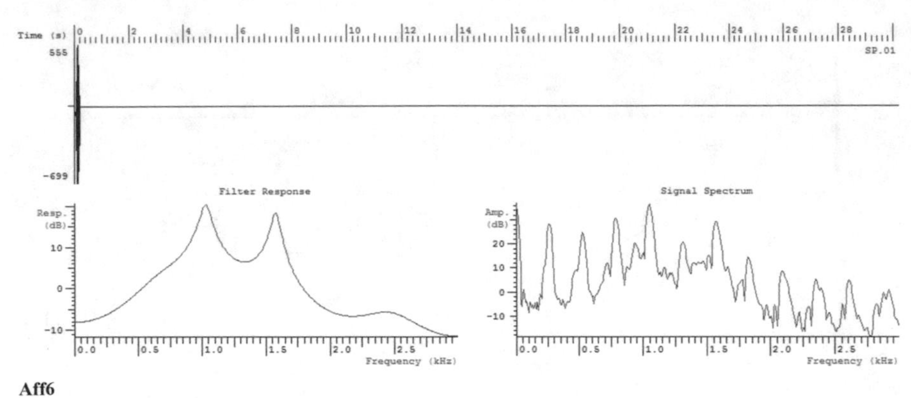

Aff6
F1 661Hz F2 1030Hz F3 1580Hz F4 2477Hz
Int: 0.1970dB
Per: 0.0001667

……………………………………………………………………………………………………

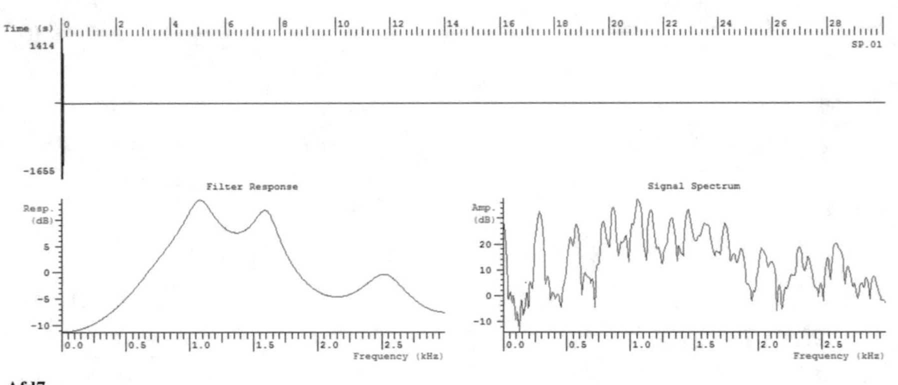

Afd7
F1 777Hz F2 1082Hz F3 1604Hz F4 2528Hz
Int: 0.0966dB
Per: 0.0001667

……………………………………………………………………………………………………

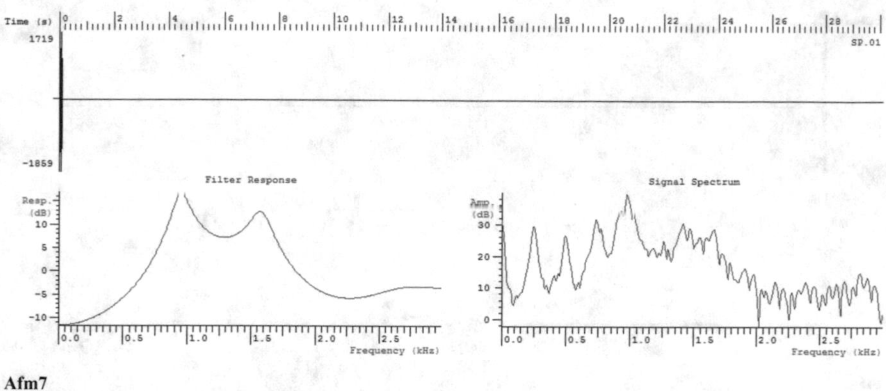

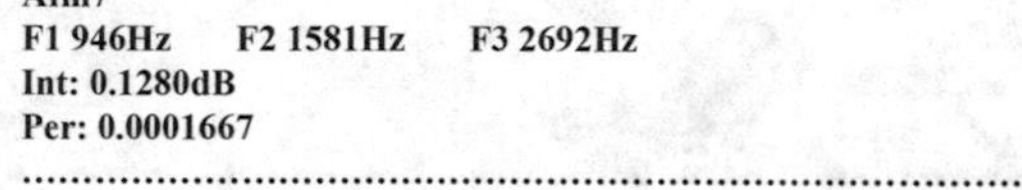

Afm7
F1 946Hz F2 1581Hz F3 2692Hz
Int: 0.1280dB
Per: 0.0001667

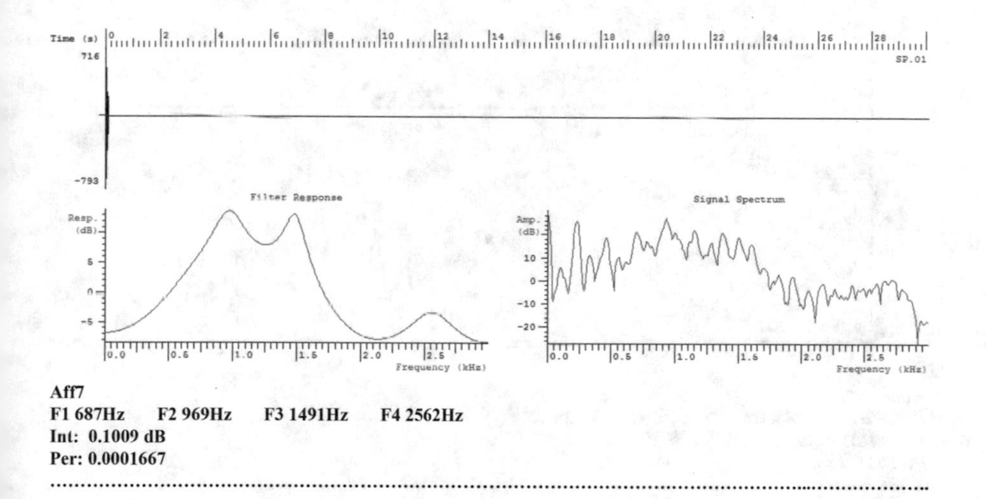

Aff7
F1 687Hz F2 969Hz F3 1491Hz F4 2562Hz
Int: 0.1009 dB
Per: 0.0001667

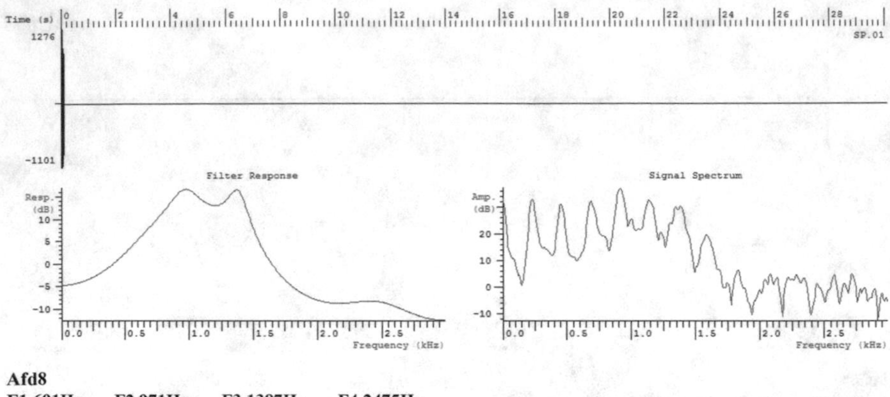

Afd8
F1 691Hz F2 971Hz F3 1387Hz F4 2475Hz
Int: 0.1280dB
Per: 0.0001667

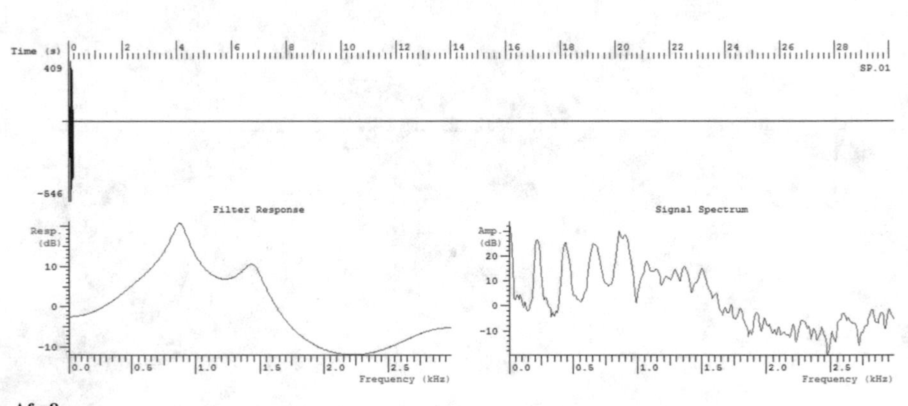

Afm8
F1 599Hz F2 880Hz F3 1448Hz F4 2815Hz
Int: 0.1632dB
Per: 0.0001667

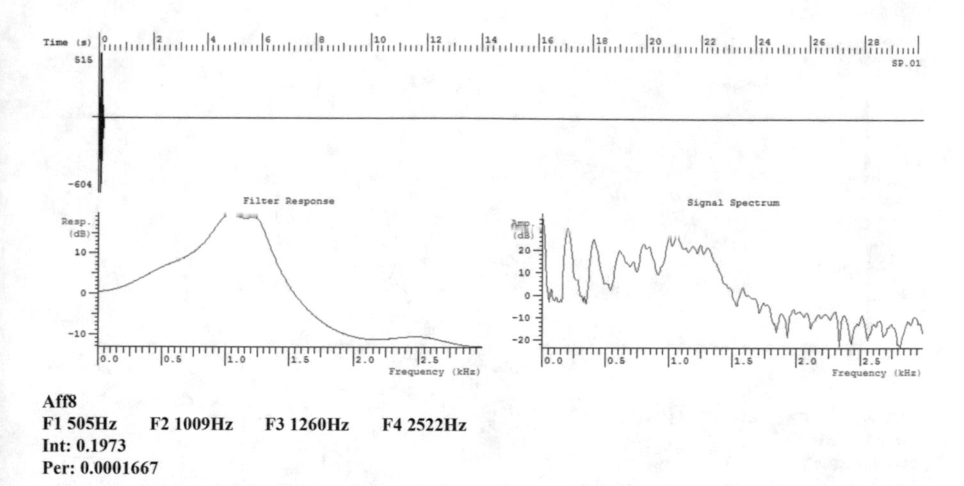

Aff8
F1 505Hz F2 1009Hz F3 1260Hz F4 2522Hz
Int: 0.1973
Per: 0.0001667

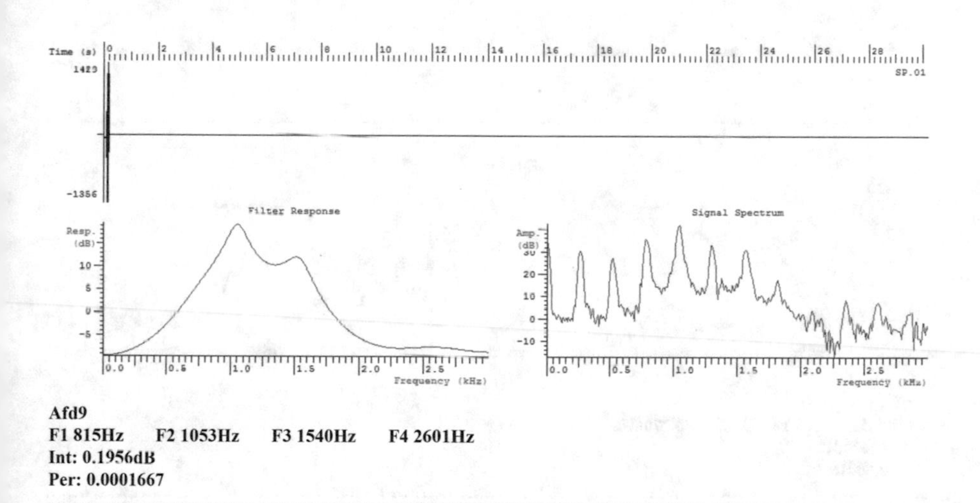

Afd9
F1 815Hz F2 1053Hz F3 1540Hz F4 2601Hz
Int: 0.1956dB
Per: 0.0001667

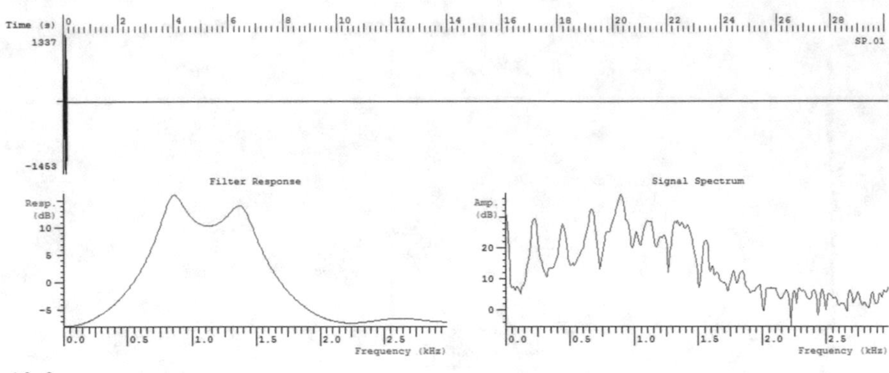

Afm9
F1 849Hz F2 1384Hz F3 2610Hz
Int: 0.1312dB
Per: 0.0001667

....

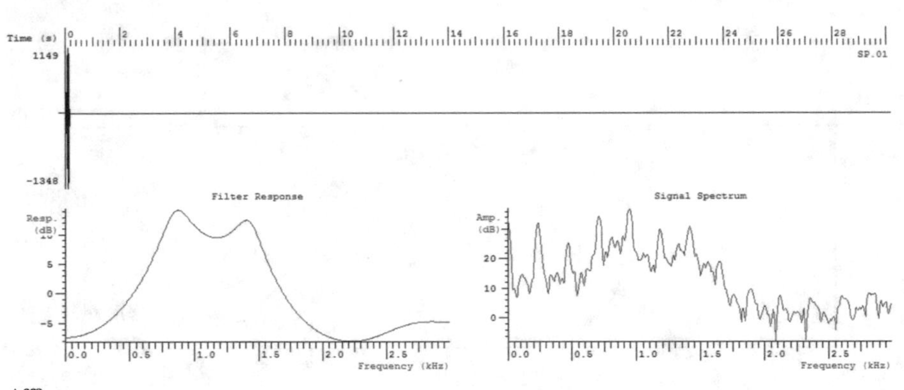

Aff9
F1 858Hz F2 1428Hz F3 2706Hz
Int: 0.1629dB
Per: 0.0001667

....

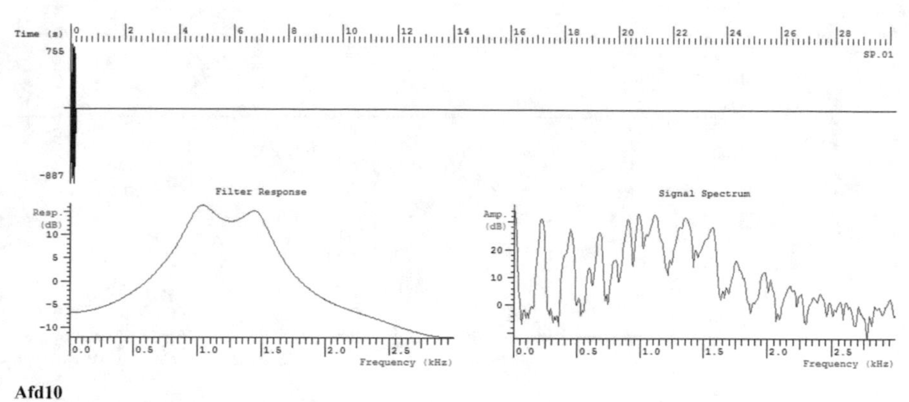

Afd10
F1 1020Hz F2 1461Hz F3 2366Hz
Int: 0.1944dB
Per: 0.0001667

....

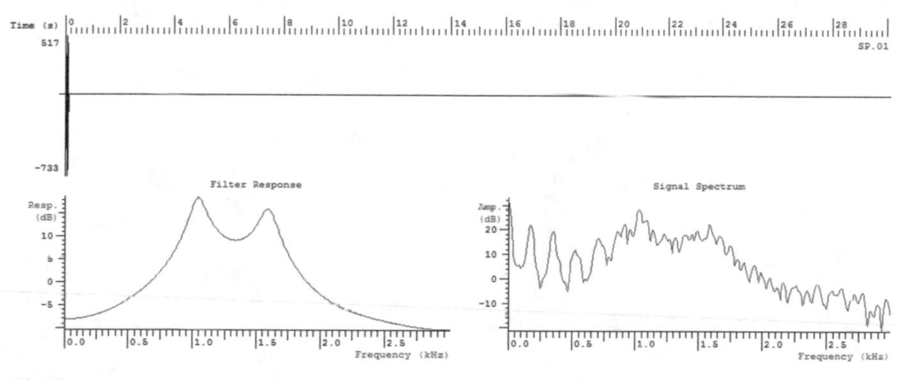

Afm10
F1 1041Hz F2 1599 Hz
Int: 0.1312dB
Per: 0.0001667

....

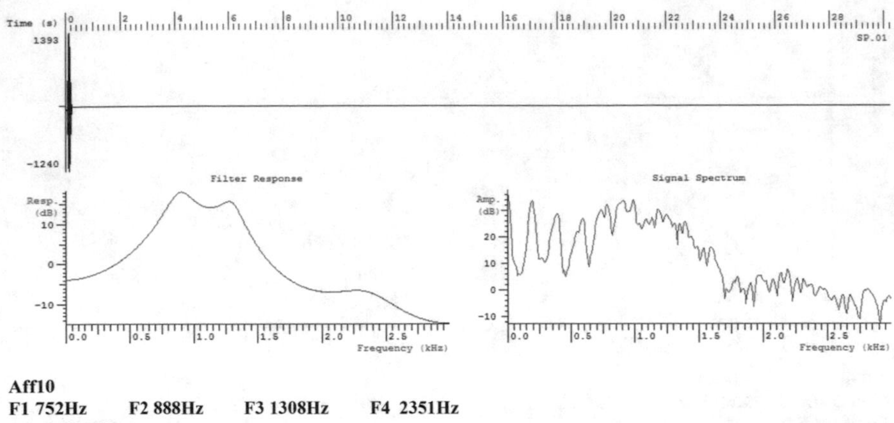

Aff10
F1 752Hz F2 888Hz F3 1308Hz F4 2351Hz
Int: 0.2278dB
Per: 0.0001667

....

Glossary

English	Arabic
Automatic recognition	التعرّف الآلي
Acoustic phonetics	الصوتيات الأكوستيكية
Aeroacoustic noiseic	ضجيج هوائي فيزيائي
After treatment	بعد المعالجة
Algorithms	الخوارزميات
Amplitude	سعة
Analog signal	إشارة تماثلية
Artificial intelligence	الذكاء الإصطناعي
Arytenoid cartilages	الغضرفان الطرجهاريان
Attitudinal function	وظيفة مقامية
Biuniqueness	ثنائية العلاقة
Coarticulation	التشكيل الصوتي
Comprehension of words	فهم الكلمات
Computing procedure	إجراء حسابي
Consonant	الصامت
Cricoid cartilage	الغضروف الدرقي
Digital signal	إشارة رقمية
Different forms of knowing	أشكال مختلفة للمعرفة
Discourse function	وظيفة خطابية
Discrimination between speech	التمييز بين الأصوات الكلامية
Fast Fourier transform	تحليل فوريي السريع
Focusing function	وظيفة تأكيدية
Frequency	تردد
Generalization	تعميم
Grammatical function	وظيفة نحوية
Infrasonic	ماتحت صوتية
Input	مدخلات

M. Dib, *Automatic Speech Recognition of Arabic Phonemes with Neural Networks*, SpringerBriefs in Applied Sciences and Technology,
https://doi.org/10.1007/978-3-319-97710-2

English	Arabic
Intelligence	الذكاء
Intensity	شدّة
Interpersonal	علائقي
Intrapersonal	ذاتي
Intonation	تنغيم
Invariance	الثابت
Linearity	سياقات خطية
Linguistic	لساني
Local determinancy	تحديد الوسط
Multimodal system	نظام متعدد الأوجه
Nasal cavity	مجرى أنفي
Neural approach	المقاربة العصبونية
Neural net works	شبكة العصبونات
Oral cavity	مجرى صوتي
Output	مخرجات
Period	دور
Periodic sound	صوت منتظم
Permanent signal	إشارة ثابتة
Pretreatment phase	مرحلة ماقبل المعالجة
Processing unit	وحدة معالجة
Production of babble	إصدار عملية البأباة
Programme	برنامج
Raw speech	كلام الأولي
Synapses	تكيهات
Sampling with replacement	الإعادةمع المعاينة
Set of connections	مجموعة من العلاقات
Simple random sample	العينة العشوائية البسيطة
Sound spectrum	الطيف الصوتي
Spectrum	طيف
Speech recognition	التعرّف على الكلام
Syllable	مقطع
Training procedure	إجراء تدريبي
Transitional signal	إشارة متخامدة
Trainability	القدرة على التمرين
Treatment	المعالجة
Ultrasonic	مافوق صوتية
Uniformity	الشكل الموحد
Vowel	الصائت

References

Atchison J (2000a) The seeds of speech language origin and evolution, Cambridge approaches to linguistics. Cambridge University Press, Cambridge. ISBN 0-521-46793-4

Atchison J (2000b) Language change progress or decay, Cambridge approach to linguistics, 3rd edn. Cambridge University Press, Cambridge. ISBN-10: 0521795354

Atchison J (2010) Linguistics teach yourself books, 5th edn. Hodder and Stoughton, London. ISBN 10: 1444105965

Berg T (2009) Linguistic structure and change an explanation from language processing, 1st edn. Routledge, New York

Burnley D (2000) The history of the English language. A source book, 2nd edn. Routledge, London. ISBN 0-582-31263-9

Cambridge Advanced Learner Dictionary (2008) 3rd edn. Legoprint. S.p.A, Italy

Cantineau J (1960) Cours de phonétique Arabe. Klincksiect, Paris

Connor OJD (1980) Better English pronunciation, 2nd edn. Cambridge University Press, Cambridge. ISBN 0-521–2315-3

Crystal D (1992) Introducing linguistics, 1st edn. Penguin English, London

Dardjowidjojo S (1972) Pros and Cons in Aila 3rd congress Copenhagen proceedings, vol. 1

Di Pietri E (1987) Regards sur le Centre Polonais d'Analyse Contrastive. Projet Contrastif français-anglais contrastes N1415

Dubois J, et groupe de linguistique, (Giacomo M, Guespin L, Marcellessi C, …) (1973) Dictionnaire de linguistique. Librairie Larousse, Paris

Dubois J, Giamo M, Guespin L, Marcelleci C, Baptistemarcelleci J, Pierre Mevel J (1994) Dictionnaire de Linguistique. Larouse. Premiére édition

Evans V, Green M (2006) Cognitive linguistics an introduction. Edinburgh University Press, Edinburgh

Ferguson C (1957) Two problems in Arabic phonology. Word 13(3)

Ferguson C (1960) Bulletin of School of African and Oriental Studies. Volume 23 Part 2 1

Fisiak J (ed) (1980) Theoretical issues in contrastive linguistics, Amsterdam studies in the theory and history of linguistic science. Series iv. Current issues in linguistic theory, vol 12. John Benjamins, Amsterdam

Fries CC (1945) Teaching and learning English as a foreign language. University of Michigan, Ann Arbor, 1945 and 1965

M. Dib, *Automatic Speech Recognition of Arabic Phonemes with Neural Networks*, SpringerBriefs in Applied Sciences and Technology,
https://doi.org/10.1007/978-3-319-97710-2

Geeraerts D, Cuyckens H (2007) The oxford hand book of cognitive linguistics. Oxford University Press, Oxford
Ginésy M (1995) Mémento de la phonétique anglaise, un outil pratique pour maîtriser parfaitement la prononciation de" l'anglais. Nathan University
Hagége C (1982) La structure des langues. 6eme édition collection fondée par Angoulvent encyclopédique
Harrap's Dictionary Easy English Dictionary (1984) la Bible des dictionnaire bilingues. Chambers Harrap, London
Hickey R (2003) Motives for language change. Cambridge University Press, Cambridge
Hornby AS (2010) Oxford advanced learner's dictionary of current English, 8th edn. Oxford University Press, Oxford
Jakobson R (1986) Essais de la linguistique générale. Les fondations du langage. Traduit par Nicolas ruwet
Jakobson R, Waugh L (1980) La charpente phonétique du language. Traduit de l'anglais par Alain. Kihm les éditions de minuit
Juangm BH, Chou W (2003) Pattern recognition in speech and language processing. CRC Press, Baton Rouge
Lieberman P, Blumstein S (1988) Speech physiology, speech perception, and acoustic phonetics, Cambridge Studies in speech science and communication. Cambridge University Press, Cambridge
Lily R, Michel V (1999) Initiation raisonnée a la phonétique de l'anglais. Hachette Superieur, Paris
Martinet A (1970) Elément de linguistique générale. Armand Colin, Paris
Martinet A (2000) Les introuvables d'André Martinet. Presses Universitaires de la France, Paris
Matthews PH (2007) Oxford concise dictionary of linguistics. Oxford University Press, Oxford
McMahon AMS (1994) Understanding language change. Cambridge University, Cambridge
Mounin G (2000) Les clefs pour la linguistique. Bibliothèque10/18 dirigé par Jean bibliothèque 10/18 dirigé par Jean Claude zylberstein
Murcia CM, Brinton DM, Goodwin MJ (1996) Teaching pronunciation. Cambridge University, Cambridge
Newmeyer FJ (ed) (1988) Linguistics. The Cambridge survey, Linguistic theory: Foundations, vol 1. Cambridge University Press, Cambridge
Richard J, Schmidt R (2010) Longman dictionary of language teaching and applied linguistics, 4th edn. Emerald, Bingley
Richard RF, Culicover PW (1997). Principles and parameters an introduction to syntactic linguistics. Oxford introductions to language study. Oxford University Press, Oxford
Siouffi G (1999) 100 fiches pour comprendre la linguistique par G.Siouffi maître de conférence à l'université de Paul Valery Montpelier
Skandera P, Brleigh P (2005) A manual of English phonetics and phonology. Narr Franke, Tübingen
Smith N (2004) Chomsky ideas and ideals, 2nd edn. Cambridge University Press, Cambridge
Sorés PLA (2008) Typologie et linguistque contrastive. Théoris et applications dans la comparaison des langues Réalization de la couverture. Thomas Jaberg, Peter Lang SA. ISBN 978 03911–518-1, ISSN 1424-3563. Imprimé en allemagne
Stemmer B, Whitaker H (2008) Handbook of the neuroscience of language, 1st edn. Academic, San Diego
The Hutchinson Encyclopedia (1999) Edition published in 27-08-1998. Helicon, Oxford. ISBN 9781859862544.
Zemmour D (2008) L'initiation à la linguistique. Chargé de cours de L'Université de Paris IV Sorbonne
Ziegler JC, Perry C, Zorzi M (2014) Modelling reading development through phonological decoding and self-teaching: implications for dyslexia. Philos Trans R Soc Lond B Biol Sci B369:20120397

Arabic References

Abdeljlil A (1998) *Handassat Al Maqatie Al Sawtiya Wa Moussiqa Al Shier*. Architecture of syllable and Music of Arabic Poetry, 1st edn. Dar Assafaa Li Al Nashr Wa Al Tawzie, Oman, Jordan

Al Faraa Y (1983) *Maani Al Quraan* (Meanings of Quran) Tahiq Mohammed Ali Al Najar wa Ahmed youcef Najati, 3rd edn (Proofread by Mohammed Ali Al Najar and Ahmed youcef Najati. Alem Al Koutoub) World of Books. Beirut, Lebonon

Al Farahidi K (1980) *Kitab Al Ayn* (The book of Al Ayn) (Proofread by Ibrahim Al Samarai and Mehdi Al Makhzoumi). Iraqi Ministry of Culture and Information

Al Jorjani A (1983) *Dalail Al Ijaz* (Proofs of incapacitation) (Proofread by Mohammed Redouane Al Daya and Fayaz Al Daya Dar Qutayba), 1st edn. Damascus, Syria

Al Mawsili MM (2005) *Maarif Atarjama Atahririya*. Ussusuha Al Nadariya wa Manahijuha Wa Falsafatuha Taraiquha Wa Tatbiquatiha Al Amaliyah (Knowledge of translation, its theoretical bases, methods, philosophy, ways, and applications), 1st edn. Beirut

Al Ouraghi M (2001a) *Al Wasait Alughawiya 1 Afoul Alissaniyat AlKoulia*, 1st edn. Dar Al Aman, Rabat

Al Ouraghi M (2001b) *Al Wasait Alughawiya Alissaniyat Alnisbiya wa Al Anhae Al Namatiya* (Linguistic parameters. General linguistics and typological grammar), 1st edn. Dar Amane, Rabat

Alakhfash S (1970) *Kitab Alqawafi* (Book of rhymes). Tahqiq Azza Hacen (Proofread by Azza Hacen publications of the direction of revitalizing the ancient patrimony), Damascus, Syria

Aljahiz, Abu Othman Amru Ibn Bahr died in 255H (2010) *Albayan wa Al Tabyeen* (The book of eloquence and oratory), 1st edn. Tahqiq Darwich Jawidi (Proofread by Darwish Jawidi. Al Maktaba Al Misriya). Egyptian Library

Almubarad M (1999) *Al Moqtadab* (Proofread by Mohammed Abdelkhalaq Adima), 4th edn. Alem Alkutub Beirut, Lebanon

Annis I (1997) *Al Aswat alughawiyah* (Speech sounds) 4th Multazamat atabaa wa anashr maktabat anglo misriya Anglo Egyptian 165 Avenue Mohammed Farid Cairo

Ashour A (1984) *Mabadiae fi Qadaya Alissaniyat Almoassira* (Principles of modern linguistic issues) (Proofread by Istambouli Rabah graduated of the University of Sorbonne, Paris, Office of University Publications Central Square Ben Aknoun, Algiers)

Ayoub A (1968) *Aswat Alugha* (Sounds of Language), 2nd edn. Cairo Al Kilani Press

Baba Amar S, Amiri B (1991). *Lissaniat Alamma fi Ilm Al Tarkeeb* (Easy general linguistics, in the science of syntax) Anwar 47 Avenue Mohamed the Fifth, Algiers

Bahnassaoui H (1991) *Al Arabiya Alfousha wa Lahajatiha* (Classical Arabic and its dialects) Kouliyat Dar Al Ouloum. fara AlFayoum. Maktabat Athaqafa Adiniya. Faculty House of Sciences. Section of Fayoum, Library of Religious Culture

Bekkouche T (1982) *Al Tasrif Al Arabi Min Khilal Ilm Al Aswat Al Hadith* (Arabic inflection from modern phonetics). Nashr wa Tawzie Abdelkrim Ibn Abdellah. Tunis. Abdelkrim Ibn Abdellah, Tunis

Brokelmann C (1977) *Fiqh Alughat Samiya* (Philology of Semantic Languages). Trans. by Ramadan Abetouwab, Professor of Linguistics in the Faculty of Letters in the University Ain Chams

Chaheen A (1980) *Al Manhaj Al Sawti Lilbiniya Al Arabiya Roueya jadida fi Al Sarf Al Arabi* (The phonetic method of Arabic structure a new vision in Arabic morphology). Muassassat Al Rissalah (Treatise Institution) Beirut

Hassan T (1974) *Manahij Albahth fi Alugha* (Research methodologies in language). Kouliat Dar alouloum Jamiat alqahira. Faculty of House of Sciences, University of Cairo, Cairo

Hassan T (1979) *Alugha Al Arabiya Maanaha Wa Mabnaha* (Arabic language its meaning and structure), House of Culture, Casablanca, Morocco

Ibrahim M (1992) *Ihyaa Al Nahw* (Revitalizing grammar), 2nd edn, Cairo

Joseph NH (2002) *Al Marjii Qamous Mouassir Arabi-Firanci* (The reference modern Arabic–French dictionary) (Proofread by Shamel Bassil) Lebanon Library, Nashiroun

Mounin G (1982) *Ilm Alugha Al Moqaran fil alqarni Al Ishrin. Tarjamat Nadjib Ghazawi* (Comparative linguistics in the twentieth century). Trans. by Nadjib Ghazawi Minister of Higher Education, Presses Alwihda Institution, Syria, Damascus)

Mubarak H (1992) *Fi Siwata Zamaniya alwaqf fi Lissaniyat alklassikiya* (Temporal phonology, Pause in classical linguistics). Dar Al Amane. Rabat

Muslim Ibn Qutayba (1981) *Taaweel Moushquil AlQuran* (Interpretation of the problem of Quran) Explained and Edited by Ahmed Saqr, 3rd edn. Dar al-Kutub al-Ilmiyah House of Scientific Books, Beirut, Lebanon

Nahr H (2007) *Ilm Dalala Al Tatbiqui fi Turat Al Arabi* (Applied semantics in the Arabic Patrimony), 1st edn. Alamal House of Publication and Distribution, Jordan

Omar AM (1977) *Mohadart fi Ilmi alugha Alhadith* (Lectures in modern linguistics), 1st edn. Kuliyat dar alouloum, Jamiat alqahira. Faculty of House of Sciences, University of Cairo

Omar AM (1997) *Dirassat Asawt Alughhawi* (The study of speech sounds) Kuliyat dar alouloum, 3rd edn. Jamiaat alqahira. Faculty of House of Sciences, University of Cairo

Quazwin K (1967) *Talkhis Al Miftah fi Al Maani wa Al Bayan wa Al Badie* (Summary of the key in meanings, eloquence and rhetoric) (Proofread by Yacin Souli died in 783H/1338). Egyptian Library

Rajihi A (1988) *Tatbeeq Al Sarfi* (Inflectional Application) dar Al Maarifa Al Jamiiya, Iskandariya. House of Academic Knowledge, Alexandria

Rumani (1976) *Rissalat Al Nakt fi Iijaz Al Quran Dimna thalath Rassail fi Iijaz Al Quran* (A treaties on Expression Accuracy in the Miracle of Quran), 3rd edn (Proofread by Mohammed Khalaf Allah Ahmed and Mohmmed Zaghloul Salam Al Maarif House, Egypt)

Sibawayhi Amru Ibn Otman Ibn Qambar (1988) *Alkitab* (Proofread by Abdassalam Mohamed Harun), 3rd edn. Alaam Alkutub, Beirut, Lebanon

Talat QZ (1998) *Alqamous Al Shamel* (Al Shamel dictionary French-Arabic) dar Al Rutab Al Jamiiya

Vendryes J (1950) *Alugha* (Language) Taareeb Abdelhamid Douakhli wa Mohammed Alqassas (Arabised by Abdelhamid Douakhli and Mohammed Alqassas) Maktabat Al Anglo Misriya Wa Matbaat Lajnat Al Bayan Al Arabi Al Qahira (Anglo Egyptian Library, Commission of Arabic Eloquence Press, Cairo)

Yaqout AS (2000) *Ilm Alugha Taqabouli* (Contrastive linguistics), 1st edn. Dar almarifa aliskandariya

Yule G (1999) *Maarifat Alugha* (Knowledge of language). Trans. by Mahmoud Feraj Abdelhafid. Dar Al wafaa Lidouniya Atibaa. Iskandariya. House of Al Wafaa of Douniya Press, Alexandria

Printed in the United States
By Bookmasters